First-Principles Approaches to Metals, Alloys, and Metallic Compounds

First-Principles Approaches to Metals, Alloys, and Metallic Compounds

Special Issue Editor

Richard Dronskowski

MDPI • Basel • Beijing • Wuhan • Barcelona • Belgrade

Special Issue Editor
Richard Dronskowski
RWTH Aachen University
Germany

Editorial Office
MDPI
St. Alban-Anlage 66
4052 Basel, Switzerland

This is a reprint of articles from the Special Issue published online in the open access journal *Metals* (ISSN 2075-4701) from 2017 to 2018 (available at: https://www.mdpi.com/journal/metals/special_issues/jz_first_principles_calculations)

For citation purposes, cite each article independently as indicated on the article page online and as indicated below:

LastName, A.A.; LastName, B.B.; LastName, C.C. Article Title. *Journal Name* **Year**, *Article Number*, Page Range.

ISBN 978-3-03897-358-4 (Pbk)
ISBN 978-3-03897-359-1 (PDF)

Cover image courtesy of Robert Gier (Steel Institute, RWTH Aachen University).

Contents

About the Special Issue Editor

Richard Dronskowski studied chemistry and physics at M ünster, and obtained his doctorate with Arndt Simon at the MPI for Solid State Research in 1990. After staying as a visiting scientist with Roald Hoffmann at Cornell, he received his habilitation at Dortmund in 1995. He then moved to RWTH Aachen, where he is Distinguished Professor and holds the Chair of Solid-State and Quantum Chemistry. His research fields comprise solid-state chemistry (e.g., carbodiimides, guanidinates, nitrides, intermetallics), neutron diffraction (e.g., POWTEX), and solid-state quantum chemistry (e.g., electronic structure, chemical bonding, LOBSTER program, thermochemistry, ab initio ORTEP). He has been Guest Professor of Tōhoku University, Director of the JARA-HPC ab initio Simulation Laboratory, and presently serves as Distinguished Chair Professor at the Hoffmann Institute of Advanced Materials in Shenzhen. Among others, he has been awarded the Otto Hahn Medal, the Prize of Angewandte Chemie, the Chemistry Lecturer Prize, the M. N. Saha Memorial Lecture, the RWTH Innovation Award, and the Egon Wiberg Lecture. He has authored Computational Chemistry of Solid State Materials (Wiley-VCH, 2005) and edited the Handbook of Solid State Chemistry (six volumes, Wiley-VCH, 2017).

metals

MDPI

Editorial

First-Principles Approaches to Metals, Alloys, and Metallic Compounds

Richard Dronskowski

Chair of Solid State and Quantum Chemistry, Institute of Inorganic Chemistry, RWTH Aachen University, D-52056 Aachen, Germany; drons@HAL9000.ac.rwth-aachen.de

Received: 4 September 2018; Accepted: 6 September 2018; Published: 7 September 2018

1. Introduction and Scope

At the beginning of the 21st century, electronic-structure theory has matured to a degree that allows for accurate phase prediction and computational characterization of various kinds of materials; in particular, elemental metals adopting whatever allotropic structure, various intermetallic compounds, and other complex metal-rich phases. Hence, fundamental theoretical progress has been made and is rapidly continuing in both physics and chemistry. From a more applied, engineering-like perspective, there is an urgent need for novel metallic structural materials, such as advanced steels, to address future challenges arising in both mechanical and civil engineering as well as energy production and conversion. While it is clear that different microstructural features influence the macroscopic behavior, modern techniques for simulation and modeling of metals and intermetallic phases at the atomic scale may enormously accelerate and guide the entire development process. In particular, atomistic understanding is a key issue because it allows for the generation of (spin-dependent) structural models of crystalline phases and the calculation of enthalpies and other free energies as a function of pressure and temperature. In combination with evolutionary algorithms and advanced thermochemical and phase-field approaches, these methods provide a solid ground for a novel methodological approach to the physics, chemistry, and engineering of metals and metal-rich materials. Furthermore, fundamental insights obtained in this manner may be incorporated, either as input parameters or key assumptions, into larger-scale models, whether purely theoretical or computational, rendering atomistic simulations essential for the development of multiscale approaches. Thus, this Special Issue focusing on first-principles approaches to metals, alloys, and metallic compounds tries to follow that train of thought, and it also aims at allowing for a wider perspective on metallic materials, to be studied by physicists, chemists and materials scientists, as well as engineers.

2. Content

To begin with, and as a timely object, high-strength high-manganese steels are at the very core of modern metal engineering, so Sevsek and Bleck [1] demonstrate an ab initio-based modelling of high-manganese steels depending on first-principles calculations of short-range ordering energies, a question of paramount importance for the Collaborative Research Centre 761 ("Steel ab initio") funded by the German Research Foundation (DFG). Both configurational structures and the impact of alloying elements are analyzed, finally providing good agreement with experimental data. In a somewhat similar manner, Song et al. [2] provide a combined small-angle neutron scattering and ab initio investigation on the Mn–C short-range ordering in an X60Mn18 steel. Not only does the experiment prove the presence of such ordering upon recrystallization, theory provides evidence for cluster formation and its evolution, which also translates into a stress-strain curve. The role of carbon, in particular carbon precipitates, is covered in the contribution by Sawada et al. [3] using the examples of titanium and niobium carbide. While the interface energy between carbide and iron is obtained via large-scale first-principles theory, the estimated coherent-semi-coherent TiC transition diameter agrees with experiment.

The aforementioned three papers already allude to bridging the gap between atomistic and continuum levels, directly covered in the contribution by Korbmacher et al. [4] who utilize Ni–H as a reasonable system to model phase equilibria. By considering various effects, they arrive at a fully quantitative agreement for the chemical potential without adjustable parameters. Likewise, Weikamp et al. [5] present a selection of scale-transfer approaches from the electronic to the continuum regime for topics relevant to hydrogen embrittlement. Eventually, they develop an approximative scheme to estimate grain-boundary energies for varying C and H contents, and they consider the dependence of hydride formation on the grain-boundary stiffness. When it comes to time evolution and dynamical phenomena, the paper by Zhang and Jiang [6] deals with molecular-dynamics simulations of crack propagation in nanoscale polycrystalline nickel. The strain rate has an important effect on the mechanism of crack propagation, and for higher strain rates local, non-3D-crystalline atoms show up, and Lomer–Cottrell locks are formed.

If we ignore nonmetallic elements for the moment and focus on intermetallic binaries, Herrig et al. [7] show how to perform low-temperature syntheses of smooth face-centered FeMn thin films provided proper guidance by ab initio theory. The latter indicates very strong interfacial bonding of the Cu nucleation layer to an alumina substrate and between fcc FeMn and Cu, hence local epitaxial growth is enabled. With respect to binary phases such as HfOs, HfIr, and HfPt, Li et al. [8] study their structural, electronic, and elastic properties using first-principles theory and confirm the order of thermodynamic stability as HfPt > HfIr > HfOs. On the other side, the calculated bulk moduli follow the order HfOs > HfIr > HfPt, and the anisotropy of acoustic velocities, Debye temperatures, and thermal conductivities are obtained.

Coming back to the critical role of hydrogen, Hüter et al. [9] present a multiscale modelling of H transport and segregation in polycrystalline steels from a chemo-mechanical model taking into account stress gradients as well as microstructural trapping sites; the energetic parameters are determined from ab initio calculations. A scale-bridging description of dislocation-induced H aggregation is accessible, but there are limitations hindering a quantitative comparison to experimental data. Likewise, Timmerscheidt et al. [10] investigate possible H-trapping effects connected to the presence of Al in the grain interior by employing density-functional theory, and they aim at understanding the relevance of short-range ordering effects because of the occurrence of Fe_3AlC κ-carbides. The individual H–H/C–H interactions are repulsive, but Mn enhances H trapping. All that can be expressed mathematically, such as to numerically describe hydrogen embrittlement. And yet, full hydrogen content bridges the gap to inorganic chemistry as shown by Gong and Shao [11] who model stability, electronic structure, and dehydrogenation of pristine and doped 2D MgH_2 from first principles. The study has implications regarding dynamical stability and dehydrogenation, and it shows that the Mn-doped system exhibits good performance for hydrogen storage and dehydrogenation kinetics.

That being said, this Special Volume includes 11 original contributions, and 7 of them deal with high-manganese steels which have come to light within CRC 761 ("Steel ab initio"). In particular, the research deals with short-range ordering from experiment and theory, and the contributions also highlight carbide-like precipitates. In addition, the authors of this volume bridge the gap between atomistic and continuum levels, in the spirit of scale-transfer approaches, in particular for hydrogen embrittlement. Then, molecular-dynamics simulations play their role in terms of crack propagation. First-principles theory is helpful for growing better intermetallic thin films, and such approaches predict structural and elastic properties of metallic binaries, too. Also, multiscale modelling of hydrogen transport is provided, and the chemical reasons for H-trapping κ-carbides are highlighted. Eventually, stability and dehydrogenation of metal hydrides are looked at. Indeed, first-principles theory has acquired a firm and supportive role in the fundamental and applied research of metals, alloys, and metallic compounds. What a wonderful evolution to witness and also to be part of!

References

1. Sevsek, S.; Bleck, W. Ab Initio-Based Modelling of the Yield Strength in High-Manganese Steels. *Metals* **2018**, *8*, 34. [CrossRef]
2. Song, W.; Bogdanovski, D.; Yildiz, A.; Houston, J.; Dronskowski, R.; Bleck, W. On the Mn–C Short-Range Ordering in a High-Strength High-Ductility Steel: Small Angle Neutron Scattering and Ab Initio Investigation. *Metals* **2018**, *8*, 44. [CrossRef]
3. Sawada, H.; Taniguchi, S.; Kawakami, K.; Ozaki, T. Transition of the Interface between Iron and Carbide Precipitate From Coherent to Semi-Coherent. *Metals* **2017**, *7*, 277. [CrossRef]
4. Korbmacher, D.; von Pezold, J.; Brinckmann, S.; Neugebauer, J.; Hüter, C.; Spatschek, R. Modeling of Phase Equilibria in Ni-H: Bridging the Atomistic with the Continuum Scale. *Metals* **2018**, *8*, 280. [CrossRef]
5. Weikamp, M.; Hüter, C.; Spatschek, R. Linking Ab Initio Data on Hydrogen and Carbon in Steel to Statistical and Continuum Descriptions. *Metals* **2018**, *8*, 219. [CrossRef]
6. Zhang, Y.; Jiang, S. Molecular Dynamics Simulation of Crack Propagation in Nanoscale Polycrystal Nickel Based on Different Strain Rates. *Metals* **2017**, *7*, 432. [CrossRef]
7. Herrig, F.; Music, D.; Völker, B.; Hans, M.; Pöllmann, P.; Ravensburg, A.; Schneider, J. Ab Initio Guided Low Temperature Synthesis Strategy for Smooth Face–Centred Cubic FeMn Thin Films. *Metals* **2018**, *8*, 384. [CrossRef]
8. Li, X.; Xia, C.; Wang, M.; Wu, Y.; Chen, D. First-Principles Investigation of Structural, Electronic and Elastic Properties of HfX (X = Os, Ir and Pt) Compounds. *Metals* **2017**, *7*, 317. [CrossRef]
9. Hüter, C.; Shanthraj, P.; McEniry, E.; Spatschek, R.; Hickel, T.; Tehranchi, A.; Guo, X.; Roters, F. Multiscale Modelling of Hydrogen Transport and Segregation in Polycrystalline Steels. *Metals* **2018**, *8*, 430. [CrossRef]
10. Timmerscheidt, T.; Dey, P.; Bogdanovski, D.; von Appen, J.; Hickel, T.; Neugebauer, J.; Dronskowski, R. The Role of κ-Carbides as Hydrogen Traps in High-Mn Steels. *Metals* **2017**, *7*, 264. [CrossRef]
11. Gong, X.; Shao, X. Stability, Electronic Structure, and Dehydrogenation Properties of Pristine and Doped 2D MgH$_2$ by the First Principles Study. *Metals* **2018**, *8*, 482. [CrossRef]

metals

[MDPI]

Article

The Role of κ-Carbides as Hydrogen Traps in High-Mn Steels

Tobias A. Timmerscheidt [1], **Poulumi Dey** [2], **Dimitri Bogdanovski** [1], **Jörg von Appen** [1], **Tilmann Hickel** [2], **Jörg Neugebauer** [2] and **Richard Dronskowski** [1,3,*]

[1]   Institute of Inorganic Chemistry, Chair of Solid-State and Quantum Chemistry, RWTH Aachen University, 52056 Aachen, Germany; tobias@totim.de (T.A.T.); dimitri.bogdanovski@ac.rwth-aachen.de (D.B.); Joerg.vonAppen@zhv.rwth-aachen.de (J.v.A.)
[2]   Max-Planck-Institute für Iron Research GmbH, 40237 Düsseldorf, Germany; dey@mpie.de (P.D.); t.hickel@mpie.de (T.H.); j.neugebauer@mpie.de (J.N.)
[3]   Jülich-Aachen Research Alliance (JARA-HPC), RWTH Aachen University, 52056 Aachen, Germany
*   Correspondence: drons@HAL9000.ac.rwth-aachen.de; Tel.: +49-241-809-3642; Fax: +49-241-809-2642

Received: 13 June 2017; Accepted: 3 July 2017; Published: 11 July 2017

Abstract: Since the addition of Al to high-Mn steels is known to reduce their sensitivity to hydrogen-induced delayed fracture, we investigate possible trapping effects connected to the presence of Al in the grain interior employing density-functional theory (DFT). The role of Al-based precipitates is also investigated to understand the relevance of short-range ordering effects. So-called $E2_1$-Fe_3AlC κ-carbides are frequently observed in Fe-Mn-Al-C alloys. Since H tends to occupy the same positions as C in these precipitates, the interaction and competition between both interstitials is also investigated via DFT-based simulations. While the individual H–H/C–H chemical interactions are generally repulsive, the tendency of interstitials to increase the lattice parameter can yield a net increase of the trapping capability. An increased Mn content is shown to enhance H trapping due to attractive short-range interactions. Favorable short-range ordering is expected to occur at the interface between an Fe matrix and the $E2_1$-Fe_3AlC κ-carbides, which is identified as a particularly attractive trapping site for H. At the same time, accumulation of H at sites of this type is observed to yield decohesion of this interface, thereby promoting fracture formation. The interplay of these effects, evident in the trapping energies at various locations and dependent on the H concentration, can be expressed mathematically, resulting in a term that describes the hydrogen embrittlement.

Keywords: steel research; κ-carbides; short-range ordering; hydrogen trapping; hydrogen embrittlement; carbide-austenite interfaces; ab initio calculations; density-functional theory

1. Introduction

Fe-Mn-Al-C quaternary alloys form an important class of modern advanced high-strength steels (AHSS). Synthetic variations of the Mn, Al, and C contents accompanied by sophisticated steel-processing techniques have led to the availability of different phases and microstructures such as ferrite (body-centered cubic, bcc), martensite (body-centered tetragonal, bct), austenite (face-centered cubic, fcc), or a mixture of those as observed in dual-phase (ferrite/martensite), duplex (ferrite/austenite), or triplex (ferrite/austenite/carbide) steels. In particular, so-called high-Mn steels, comprising a broad spectrum of Fe-based alloys with Mn contents in the range of 20–30 wt % and Al and C contents in the ranges of 0–10 wt % and 0–2 wt %, respectively, have attracted attention [1]. These steels exhibit particular deformation mechanisms, such as transformation-induced plasticity (TRIP), twinning-induced plasticity (TWIP), or microband-induced plasticity (MBIP) [2], thereby yielding extraordinarily high strength and ductility. Steels containing considerable amounts of Al in addition to Mn have raised recent interest due to the precipitation of ordered carbides, both from ferritic and austenitic matrices [1]. These κ-carbides have the stoichiometry Fe$_3$AlC and crystallize in the perovskite structure type (also commonly labeled as $E2_1$ or $L1'_2$) as visualized in Figure 1. Such phases have been shown to have an additional strengthening effect on these steels [3–5].

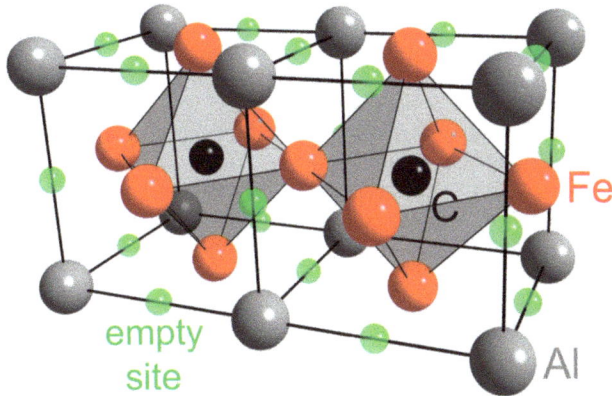

Figure 1. Crystal structure of Fe$_3$AlC, a κ-carbide crystallizing in the $E2_1$ type, shown as a 2 × 1 × 1 supercell. Grey, red, and black spheres denote Al, Fe, and C atoms, respectively. Green translucent spheres indicate the unoccupied octahedral sublattice sites.

A critical aspect with respect to the application of high-Mn steels is their large susceptibility to hydrogen-induced delayed fracture (HIDF) [6]. Hydrogen, which is in contact with steels during the production and/or the application process, may dissolve into the material, followed by migration and preferred enrichment in critical regions, such as grain boundaries or dislocations. As a consequence, local embrittlement occurs, which can finally lead to crack formation and propagation, resulting in eventual failure of the material. It has been empirically observed that the presence of small amounts of Al lowers the tendency for hydrogen embrittlement [7]. Complete prevention, however, cannot be ensured. While an alleviation of local stress due to a change in stacking fault energy (SFE) with Al inclusion has been proposed as one possible factor [8], this behavior may also be attributed to the presence of homogeneously distributed nano-sized κ-carbides or short-range ordering of similar atomic configurations which can serve as H traps, often at a precipitate/matrix interface.

For instance, experimental studies using atom probe tomography have demonstrated segregation of hydrogen at an interface of TiC with a ferritic matrix [9]. This was later confirmed by first-principles calculations [10], which further demonstrated that H accumulation can weaken the interface, resulting in hydrogen-enhanced decohesion and subsequent fracture formation.

In the case of κ-carbides it is interesting to note that, in real materials where they are formed as precipitates, their composition (Fe,Mn)$_{3+y}$Al$_{1-y}$C$_x$ may significantly deviate from the ideal stoichiometry (Fe,Mn)$_3$AlC of phase-pure κ-carbides such as the (nearly ideal) ternary Mn-rich carbide studied in a previous work [11]. For κ precipitates, the Al content depends on its concentration in the bulk phase, which leads to different values reported in the literature. For example, Andryushchenko et al. [12] quoted a range of $-0.2 < y < 0.2$, whereas Seol et al. [13] measured much lower Al contents with $y \approx 0.6$. In recent works [14,15], the Al depletion in κ-carbide is correlated with the reduced C content in these precipitates. The latter varies a lot and is typically between $0.4 < x < 0.72$ [13,16]. We explained the C off-stoichiometry as a compromise between the gain in chemical energy during partitioning and the elastic strains emerging in coherent microstructures. This off-stoichiometry is relevant for the interaction with H, because the vacant C positions are favorable trapping sites.

Traps are sometimes designated beneficial or benign [17] because they may capture diffusive hydrogen and delay its further migration to malign traps causing HIDF. By definition, the trapping energy E_{trap} is the capability of a microstructure feature to bind hydrogen better than the bulk matrix. The usual experimental approach to determine E_{trap} is thermal desorption spectroscopy (TDS) after galvanostatic or potentiostatic charging of the specimen with hydrogen. The relatively large scattering in the E_{trap} values collected from different experiments can be attributed to several causes. One particular problem is the absence of a definite reference. Measured trapping energies are often quoted relative to solution enthalpies of hydrogen in the respective bulk matrix. Since these depend on the crystal structure and the composition, a common reference state would ensure a better comparability of trapping energies from different studies.

Within theoretical studies, trapping energies have been calculated for various materials. In the case of Fe, the relevance of substitutional transition-metal atoms [18], vacancies [19], grain boundaries [20] and some precipitate phases [10] have been discussed. To the best of our knowledge, however, the hydrogen trapping by κ-phases has not been explicitly addressed theoretically. There are two experimental publications which assume an irreversible hydrogen trapping at the interface between κ-carbides and the Fe matrix [21,22], and another contribution attributing rather large activation energies of 76 and 80 kJ/mol for hydrogen desorption in the κ-carbides [23]. The mechanistic details, however, are not yet understood.

In the present work, the capability of κ-carbides to bind hydrogen is investigated using ab initio electronic-structure techniques within the framework of density-functional theory. We will analyze the role of C vacancies, as well as that of the interface between κ-carbide and Fe matrix in trapping H. In order to reveal the chemical nature of this trapping, the C–H and H–H interactions in the $L1_2$-Fe$_3$Al phase are assessed. Further, the influence of manganese substitution on the trapping capability is calculated. Finally, the suitability of the κ-carbide/austenitic Fe matrix interface as a hydrogen trap is investigated, followed by the study of enhanced decohesion caused by the presence of hydrogen at the interface.

2. Computational Methods

All quantum-mechanical structure optimizations and total-energy calculations were performed with the Vienna ab initio Simulation Package (VASP, version 5.4.1, Computational Materials Physics, University of Vienna, Vienna, Austria, 2015) [24–27], a software package employing density-functional theory (DFT) and utilizing plane waves together with PAW/pseudopotentials, thereby especially suited for simulations of periodic systems. The crystal orbitals were expanded in plane waves by means of the projector-augmented wave method (PAW) [28,29] with a kinetic energy cutoff of 500 eV. Contributions from exchange and correlation interactions were approximated utilizing the generalized-gradient approximation (GGA) functional in the parametrization by Perdew, Burke, and Ernzerhof [30]. Brillouin zone integration was performed by the scheme of Monkhorst and Pack [31] using **k**-point grids of $n \times n \times n$ with $n = 12$, 8, 4, and 4 for the four-atom (stoichiometric compounds), 32-atom ($2 \times 2 \times 2$ fcc supercell), 108-atom ($3 \times 3 \times 3$ fcc supercell), and 128-atom ($4 \times 4 \times 4$ bcc supercell) metal lattices, respectively. The interface (between κ-carbide and fcc-matrix) calculations were performed for a 92-atom ($2 \times 2 \times 5$ supercell) and 108-atom ($2 \times 2 \times 6$ supercell) system, where the total number of atoms varies with the thickness of the κ-carbide. The corresponding **k**-point grids employed were $8 \times 8 \times 5$ and $8 \times 8 \times 3$, respectively. Partial band occupancies were considered using the smearing scheme of Methfessel and Paxton [32] with the σ value set to 0.2 eV. The magnetic properties of the bulk κ-phases were taken into account by performing spin-polarized computations assuming a ferromagnetic model; nonetheless, this was restricted to ordered collinear magnetism. Non-collinearities are known to reduce the ground state energy in Fe-Mn alloys only slightly and will have little relevance at finite temperatures [33]. All other calculations, including interface models, were performed for the non-magnetic case, unless otherwise indicated.

For ideal κ-structures, the formation energies were calculated using:

$$\Delta E_{el} = E_{el}(\text{Fe}_3\text{AlX}) - [E_{el}(\text{fcc-Al}) + 3\,E_{el}(\text{fcc-Fe}) + E_{el}(\text{X})] \tag{1}$$

where $E_{el}(X)$ is the ground-state energy of graphite (X = C) or half the energy of an H_2 molecule (X = H). We note that ΔE_{el} is a theoretical energy, often provided in the ab initio community and used here solely for comparisons with earlier contributions in other studies. In order to evaluate precipitation behavior, not pure reference materials but the chemical potentials of the components as determined by the matrix materials are relevant. For both educts, the fcc structure was used as a reference, being the stable phase of Al and high-Mn steels, and forming a coherent interface with the perovskite-type κ-carbide.

The trapping energies were determined by the solution enthalpy (corresponding to the local chemical potential) of a hydrogen atom on the site in question as compared to an H atom in a γ-Fe (fcc) matrix, which is thus the reference state:

$$E_{trap} = [E_{el}(\kappa - Al_8Fe_{24}C_xH) - E_{el}(\kappa - Al_8Fe_{24}C_x)] - [E_{el}(fcc\text{-}Fe_{108}H) - E_{el}(fcc\text{-}Fe_{108})] \quad (2)$$

For determining the H solubility at a given site, (zero-point) vibrations can be important [34]. Our study, however, is primarily concerned with trapping energies of hydrogen, i.e., the difference in the solution enthalpy of hydrogen for different sites and regions of steel materials comprising ferritic and austenitic metal matrices, carbide precipitates, as well as the carbide/matrix interface. While the (zero-point) vibrational energies of hydrogen in a gas phase of H_2 molecules and of interstitial hydrogen in a metal matrix differ substantially, the difference is negligible when various interstitial sites are compared with each other. Thus, for the remainder of our study, the total energy of all calculated structures is approximated as their DFT energy.

The choice of the reference state in Equation (2) does not alter the conclusions if different positions and configurations within the same trapping phase are compared, which is the primary scope of this work. Furthermore, the potential of a phase to trap diffusive hydrogen with respect to other metal matrices than γ-Fe (fcc) can be obtained by adding the trapping energy for such a matrix with respect to γ-Fe. Determining the trapping energy of solid solutions such as Al-containing ferrite or Mn-containing austenite can, however, be exhaustive since a very large number of possible hydrogen positions would have to be considered due to the statistically random, i.e., disordered, nature of these alloys.

Similarly to Equation (2), the hydrogen solution enthalpy in the interface was computed using the following expression:

$$\begin{aligned} E_{trap}(int) = &[E_{el}(H @ \kappa\text{-}Al_{12}Fe_{28}C_{12}/fcc\text{-}Fe_{56}) - E_{el}(\kappa\text{-}Al_{12}Fe_{28}C_{12}/fcc\text{-}Fe_{56})] \\ &- [E_{el}(fcc\text{-}Fe_{108}H) - E_{el}(fcc\text{-}Fe_{108})] \end{aligned} \quad (3)$$

where the first two electronic energies on the right-hand side of the expression correspond to the fully-relaxed energies of the supercell comprised of κ-carbide and fcc-Fe with and without hydrogen, respectively.

3. Results and Discussion

3.1. Stoichiometric Phases of L1$_2$ and E2$_1$ Symmetry

The formation of κ-carbide requires the simultaneous partitioning of substitutional Al and interstitial C from the solid solution. Here, we focus on the competing C and H interstitials and consider their energetics in Fe_3Al. Although we cannot confirm a previous first-principles result that the $L1_2$ structure is the ground state of Fe_3Al [35], but obtain DO_3 instead (Table 1), we will use $L1_2$ as a matrix phase in the upcoming considerations. This is because the small energy difference of approx. 1 kJ/mol is compensated by the stabilization effect due to C insertion, yielding the perovskite $E2_1$-Fe_3AlC, commonly known as κ-carbide (Figure 1). Our calculations of the lattice parameter, total energy, and the local magnetic moments as given in Table 1 agree well with other calculations [35–37].

In particular, we confirmed earlier findings [15] that the formation of this phase out of pure elements is exothermic.

Table 1. Lattice parameter a, magnetic saturation moments μ_{theo} and formation energies ΔE_{el} at $T = 0$ K of various compounds related to the κ-phase.

Compound	Structure	a_0 (Å)	μ_{theo} (μ_B/Fe atom)	ΔE_{el} (kJ/mol)
Fe_3Al	BiF_3, $D0_3$	5.74	2.0	−78
Fe_3Al	Cu_3Au, $L1_2$	3.65	2.3	−77
Fe_3AlC	perovskite, $E2_1$	3.75	1.0	−90
Fe_3AlH	perovskite, $E2_1$	3.68	2.1	−85

When investigating the relevance of these κ-carbides as a trap for H, we first consider the ideal phase with the composition Fe_3AlC. The calculations confirm the finding for pure fcc-Fe [38] that the empty octahedral sites (green spheres in Figure 1) are more favorable for H ($E_{trap} = 0.13$ eV) than the tetrahedral ones ($E_{trap} = 0.75$ eV). Nonetheless, the absolute value is endothermic, i.e., H would not enter this phase. The reason for this repulsion can be the local atomic configuration with two Al atoms in nearest-neighbor positions (first coordination sphere, CS) and/or the presence of C in the adjacent octahedral sublattices.

In order to confirm the impact of the metallic atoms, we have investigated the Al–H interaction in an fcc-Fe matrix (Figure 2). Considering H as the central atom, the interaction turns out to be repulsive if Al is located in the first CS, but an energetic optimum exists if Al and H are second-nearest neighbors, Al being in the second CS. While the first configuration matches the H position in the empty octahedral sublattice sites of κ-carbide, the second corresponds to an H atom replacing a C atom on the central interstitial site (green and black spheres in Figure 1, respectively). Thus, the central interstitial site is energetically preferred for H incorporation while the non-central octahedral sites are generally assumed to be empty and, thus, not considered in the numbering of CS.

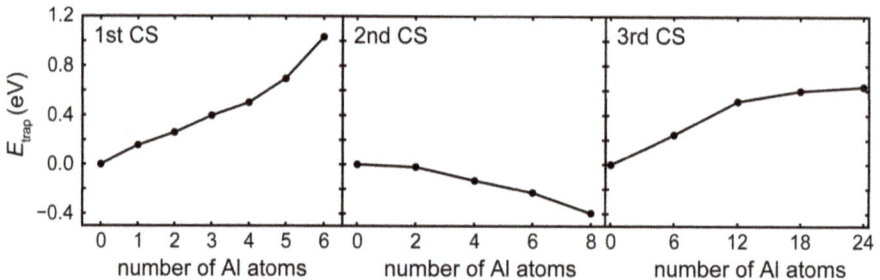

Figure 2. Hydrogen trapping energy E_{trap} as a function of the local atomic environment with respect to the amount of Al atoms in the first, second, or third coordination sphere (CS) around a single interstitial H atom in a $3 \times 3 \times 3$ supercell of fcc-Fe. Al atoms occupy Fe sites in the host matrix. The energies are defined according to Equation (2).

The attractive interaction between a single Al and a single H atom in the second CS lets one expect that the central (body-centered) interstitial sites in an $L1_2$-Fe_3Al matrix could be completely occupied by H. This would result in an $E2_1$-Fe_3AlH phase (κ-hydride), which has not been observed experimentally and has also been identified as chemically unstable in a theoretical contribution [35]. In contrast to said contribution, we obtain that the formation of this phase out of the pure elements according to Equation (1) is exothermic (−85 kJ/mol). In such a hypothetical $E2_1$-Fe_3AlH phase, the effects of H on the κ-phase formation are qualitatively similar to those of C, but quantitatively much smaller: the lattice parameter increases by 0.8% from 3.65 to 3.68 Å (2.7% increase for C), and the local magnetic moments of the Fe atoms lower by 9% from 2.3 to 2.1 μ_B (57% for C).

We note, however, that C also prefers a local coordination with Al atoms in the second CS, as shown in previous studies [39,40]. There is, therefore, a competition between the interstitial elements C and H for the formation of the κ-phase. At $T = 0$ K, the formation energy of the κ-carbide Fe_3AlC is slightly larger (−90 kJ/mol) than the one of κ-hydride (−85 kJ/mol). More important is the fact, however, that H already becomes too mobile to remain in the steel matrix before those temperatures are reached at which Al can form locally-ordered structures (approx. 600 °C for typical annealing times). We can, therefore, expect that the $E2_1$ phase is primarily formed as κ-carbide.

3.2. Competition of H and C in the $L1_2$ and $E2_1$ Structure

Despite the preference of κ-carbide formation, the typically observed C depletion in this phase yields various vacancies on its sublattice, i.e., potential sites for H that are in a favorable configuration with respect to Al. To further elaborate on this, the average trapping energy per H atom in a fully filled κ-matrix was calculated as a function of the C concentration. Carbon concentrations of 37.5–75 at % were taken into account in accordance with the typically experimentally observed C contents of Fe-rich κ-carbide precipitates. The distribution of C atoms among the different positions (central interstitial sites, referred to as κ positions) was random, while the remaining κ positions were occupied by H atoms. The calculations were performed employing 3 × 3 × 3 supercells with full structural optimization with respect to shape and volume. The results show, surprisingly, that the average trapping energy per H atom is largely independent of the C concentration (Figure 3, purple, uppermost curve).

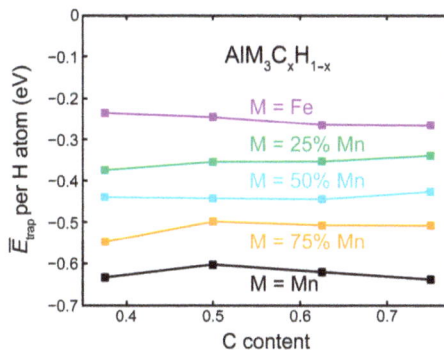

Figure 3. Average trapping energy $\bar{E}_{trap}$ per H atom as a function of the C content for a 3 × 3 × 3 $E2_1$-(Fe,Mn)$_3$Al(C,H) supercell with all κ positions filled and considering different Mn contents.

The calculations in Figure 3 have also been performed for different Mn contents, the reason being that κ-carbides growing from Mn-rich austenitic matrices contain high amounts of manganese as a substituent for iron. For this purpose, the aforementioned calculations were repeated for a 3 × 3 × 3 $E2_1$-(Fe,Mn)$_3$Al(C,H) supercell with randomly-distributed Mn atoms. Upon introducing manganese, no change can be observed with regard to the average trapping energy being independent of the carbon content (Figure 3).

The trapping capability, however, increases (i.e., the trapping energy is lowered) with increasing Mn content. This is consistent with previous findings that the short-range Mn–H interaction in austenitic alloys is attractive [38]. To study said interaction in κ-carbides, a 3 × 3 × 3 $L1_2$-Fe$_{1.5}$Mn$_{1.5}$Al supercell was used, where the Mn and Fe atoms were distributed randomly. The H atom was placed on κ positions with different numbers of Mn atoms (positions with 1–5 Mn atoms are present in the given supercell) in the surrounding octahedron. Our results show an almost linear dependence of the trapping energy on this number (Figure 4), confirming that previous findings apply to Al-containing systems as well. However, the results do not fully capture the changes of trapping energies with increased Mn content as seen in Figure 3. An important reason is that the data points in Figure 4 (for a fixed C content) all correspond to the same volume, while each Mn concentration in Figure 3 has its own equilibrium volume.

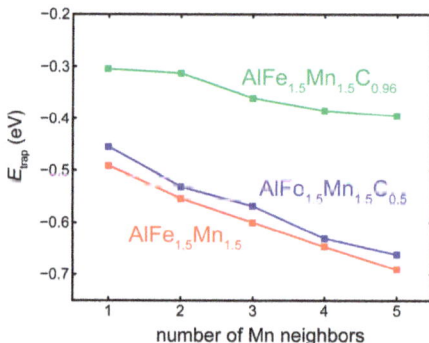

Figure 4. Trapping energy E_{trap} of a single H atom as a function of the number of Mn atoms in its first CS for a $3 \times 3 \times 3$ $L1_2$-$Fe_{1.5}Mn_{1.5}Al$ supercell with C sublattice concentrations of 0% (red), 50% (blue), and 96% (green).

To understand this effect, the trapping energy was calculated as a function of the lattice parameter (Figure 5). The calculation was performed for the $L1_2$ configuration of Fe_3Al, which has an equilibrium lattice parameter of $a(L1_2) = 3.65$ Å as well as pure fcc-Fe with $a(\gamma\text{-Fe}) = 3.45$ Å. The matrices consisted of 32-atom $2 \times 2 \times 2$ supercells. In the $L1_2$-Fe_3Al matrix a single H atom was again placed on an octahedral sublattice site normally occupied by C in the κ-phase. The calculations were performed for two different magnetic states. In all cases, the trapping capability increases with the volume. For the nonmagnetic state the H atom clearly favors the $L1_2$-matrix over pure fcc-Fe at all lattice parameters and, thus, volumes. If magnetism is considered, the ferromagnetic $L1_2$ matrix is, again, favored, apart from the special case when its volume is identical to the equilibrium volume of pure fcc-Fe (3.45 Å). Comparing both phases at the same lattice parameter, however, is only relevant in the case of a perfectly coherent interface between the matrix and the precipitate phase. Otherwise, the larger lattice parameter of the $L1_2$ phase will always yield a lower E_{trap}.

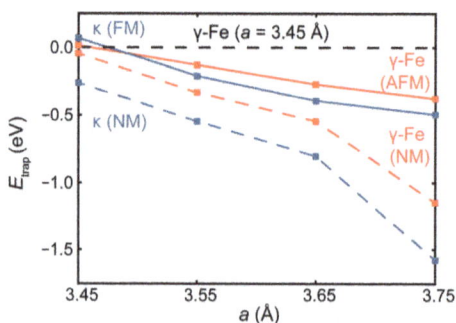

Figure 5. Trapping energy E_{trap} of a single H atom as a function of cell volume (via the lattice parameter a) for fcc-Fe (red) and the $L1_2$-Fe_3Al matrix (blue) considering different magnetic states, with the dashed/solid lines signifying the nonmagnetic/magnetic cases.

Additionally, the influence of carbon atoms on the Mn–H interaction was investigated (Figure 4). For this purpose, the supercell offering 27 κ positions was filled with 0, 14, or 26 C atoms (corresponding to a C sublattice concentration of 0, 50, or 96%), avoiding nearest-neighbor C–H configurations in the case of 14 C atoms. We observe that the $E(x_{Mn})$ linearity is independent from the C content although the attractive interaction between Mn and H seems to be reduced in the 96% C case. More importantly, however, one observes a small shift in the trapping energy if half the cell's κ positions are filled with carbon, but a significantly larger shift if the entirety of these positions is filled, the latter being attributable to the interaction between direct C–H neighbors. One may, therefore, expect that the interaction of H with C affects the trapping energy, and this raises the question why such an effect is not observed in Figure 3.

In the next step we, therefore, investigated the interaction of H with other C and H atoms (Figure 6). For this purpose, we chose $3 \times 3 \times 3$ $L1_2$-Fe$_3$Al supercells. The trapping energy of a hydrogen atom was first calculated as a function of the number of neighboring C atoms in the third CS (being the nearest neighboring κ positions if other octahedral sites are unoccupied, labeled NN), keeping the shape and volume of the supercells fixed (Figure 6a). Similar to Figure 4, we observe a repulsive effect which results in an almost linear increase of E_{trap}. Furthermore, the trapping energy depends on the configuration of the neighboring C atoms. This observation, however, is partially due to the choice of the supercell: two C atoms in the *trans* isomer case are located in neighboring unit cells of the $3 \times 3 \times 3$ supercell (energy difference to the *cis* isomer is 0.06 eV), while this is not the case in a 144-atom $4 \times 3 \times 3$ $L1_2$-Fe$_3$Al supercell (respective energy difference is 0.04 eV).

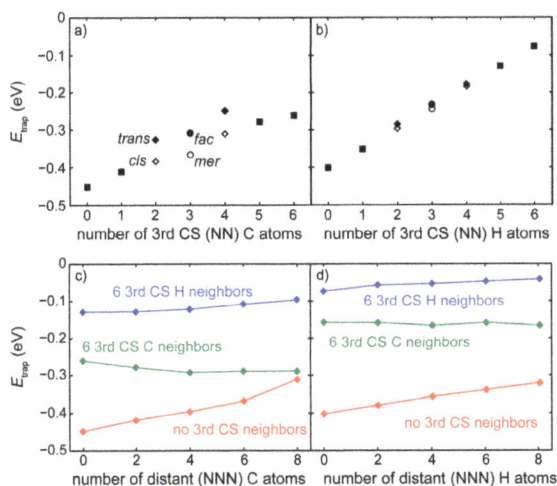

Figure 6. Trapping energies E_{trap} for a single H atom in the central κ position as a function of the number of C or H atoms in various κ positions in a $3 \times 3 \times 3$ $L1_2$-Fe$_3$Al supercell. (a) C atoms are present as nearest neighbors for H in the third CS (nearest κ neighbors, NN), under consideration of structural isomers. The lattice parameter is a = 3.69 Å. Squares signify arrangements without structural isomers with zero, one, five, or six neighbors; open/closed diamonds signify the two isomers for two neighbors (*cis* and *trans*); open/closed circles signify the two arrangements for three neighbors (*mer* and *fac*). (b) As before but for H atoms in the third CS. Energetic differences between structural isomers are negligible in this case. The lattice parameter is a = 3.65 Å. (c) C atoms are next-nearest neighbors (NNN) located at the cell corners (d_{C-C} = 6.4 Å), with no atoms (red diamonds), six H atoms (blue diamonds) or six C atoms (green diamonds) located simultaneously in the third CS (NN) κ positions, respectively. Lattice parameter as in (a). (d) As in (c), but considering H atoms as NNN neighbors. Lattice parameter as in (b). The difference in a explains the slight energetic difference in the case of zero NN/NNN neighbors.

We further note that the C–H interaction for H on a κ position and C in the non-central interstitial site coplanar to Al is attractive (not shown). Concerning H–H interactions, we observe a similar, almost linear dependence of E_{trap} on the number of neighboring H atoms in the third CS (Figure 6b). This trend is (up to four third CS H atoms) almost identical to the C–H case and has a significantly lower dependence on the atomic configuration (rendering a distinction between the various configurations unnecessary). The similarity of C–C and C–H interactions explains the constant behavior in Figure 3, since the number of third CS H and C atoms is always six in this figure.

The impact of neighboring C atoms was compared to that of C atoms in other shells. In Figure 6c the situation is exemplified for C atoms in next-nearest-neighbor (NNN) positions (corners of a $3 \times 3 \times 3$ $L1_2$-Fe$_3$Al supercell). Despite a rather long distance of $\sqrt{3} \times 3.69$ Å = 6.4 Å, a repulsive effect is still observed (red diamonds). The slope of the C dependence, however, is smaller than for the third CS neighbors by a factor of ≈2. It is further observed that the dependence on the number of NNN carbon neighbors is strongly suppressed if the NN κ positions are occupied either by C or by H. These trends can be understood as resulting from local relaxations. They indicate, once again, that the nearest κ positions are most relevant for the value of the H solubility. The trends for distant H–H interactions are, as in the case of the third CS interactions, qualitatively identical to the C–H case, as seen in Figure 6d.

3.3. H at Interfaces between κ-Carbide and the fcc-Fe Matrix

The insights obtained for the Al–H and C–H interactions are used to understand chemical trends for incorporating hydrogen in the (100) interface between κ-carbide and the fcc-Fe matrix. In this case, however, we limit ourselves to a single H atom. Various scenarios for said incorporation are possible. We first note that stoichiometric κ-carbide has two kinds of (100) layers, one containing Al atoms in one of the fcc sublattices, the other containing C atoms in the central interstitial κ positions. The considered layers for the representation of the κ-carbide in Figure 7 are indicated by a stacking sequence, e.g., C*–Al*–C*–Al*–C*–Al*. The termination layer, which can contain C or Al, is considered as the interface plane.

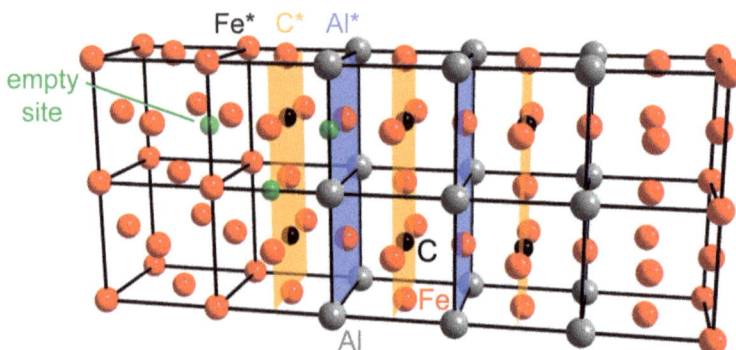

Figure 7. The $5 \times 2 \times 1$ supercell composed of κ-carbide and fcc-Fe (assuming coherent interfaces) for a given thickness of the κ-phase. Orange planes, labeled C*, signify (100) layers containing C and Fe atoms, while blue planes, labeled Al*, contain Fe and Al. Layers containing Fe only are unmarked and labeled Fe*. The translucent green spheres represent empty octahedral sites which (in addition to C vacancies) may incorporate H atoms.

The results for three different supercells, as given in Table 2, can be interpreted as follows: on the one hand, H tries to avoid positions with Al as closest neighbor in the first CS (compare with Figure 2), so incorporation into an Al-containing layer is unfavorable. On the other hand, a configuration with C in a second CS position (in contrast to the NN site shown in Figure 6a) yields an improved trapping behavior. The non-central octahedral sites in the C layer in a Fe–C–Al scenario (green sphere in orange-colored C* layer, Figure 7) are, therefore, most attractive for H, because they allow for four second CS neighbor C atoms, but only one first CS neighbor Al atom. Only slightly less favorable are the octahedral sites in Fe layers adjacent to a C-terminated κ-carbide (green sphere in Fe* layer, Figure 7), because in this case two second CS C atoms are available, but no first CS Al atoms are present. These configurations already yield a trapping at the interface.

Table 2. Hydrogen trapping energies E_{trap} for octahedral sites at the interface layer between κ-carbide and fcc-Fe, or one of the adjacent layers. Three kinds of stacking sequences in a $2 \times 2 \times 5$ supercell (SC) are provided, where layers used for H incorporation are highlighted in bold face.

Stacking Sequence of Layers in SC	E_{trap} (H) Adjacent Fe Layer (eV)	E_{trap} (H) Interface Layer (eV)	E_{trap} (H) Adjacent κ Layer (eV)
Fe*–**C***–Al*–C*–Fe*–Fe*–Fe*–Fe*–Fe*–Fe*	−0.20	−0.18 (C)	0.04 (Al)
Fe*–**Al***–C*–Al*–C*–Fe*–Fe*–Fe*–Fe*–Fe*	0.01	0.003 (Al)	−0.05 (C)
Fe*–**C***–Al*–C*–Al*–C*–Al*–Fe*–Fe*–Fe*	−0.20	−0.24 (C)	0.04 (Al)

Competing with the H incorporation discussed so far, the replacement of a C atom by an H atom should also be considered for the interface. To understand how efficiently C vacancies act as hydrogen traps near the interface, we also compare the vacancy formation energy at different positions (Figure 8). This energy amounts to 1.65 eV for a vacancy in the bulk of the κ-carbide phase, is almost the same (1.75 eV) for C atoms next to Al-terminated interfaces, but is significantly lower (approx. 0.75 eV) for C atoms at C-terminated interfaces. This is not surprising, since these C atoms have a lower number of Al neighbors in the second CS.

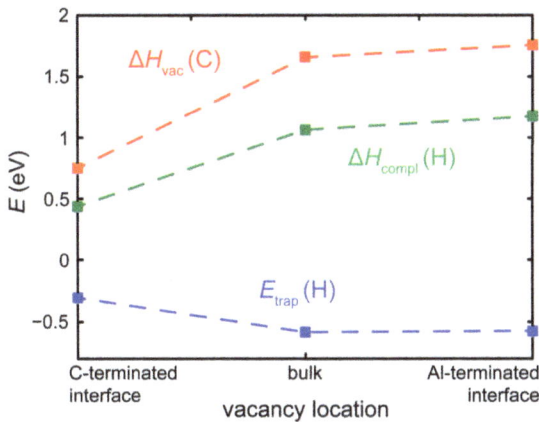

Figure 8. C vacancy formation energy (shown in red), H trapping energy within the vacancy (shown in blue) and their combination, i.e., the H–vacancy complex formation energy, (shown in green) for three different positions of vacancies. Note that trapping of H by vacancies near C-terminated interfaces is energetically the most favorable.

The differences in H trapping energies for the three different vacancy positions are less pronounced. For C vacancies in the bulk of the κ-carbide, H has a favorable configuration with respect to the Al atoms, but an unfavorable configuration with respect to the other C atoms. The resulting trapping energy is −0.59 eV. (Note that this data refers to non-magnetic calculations, so that the absolute values cannot be compared directly with Figure 4.) The situation is almost unchanged (−0.58 eV) for the C positions next to Al-terminated interfaces. If H is incorporated into C vacancies in a C-terminated interface, then both the number of Al neighbors in the second CS and the number of C atoms in NN positions are reduced from 8 to 4 and from 6 to 5, respectively. Since the change in the number of Al neighbors is more significant, the trapping energy for H in these sites is reduced (-0.31 eV).

It should be noted, however, that C vacancies which are more likely to form near the interface are more accessible, at least kinetically, to H atoms trapped near the interface than vacancies formed in the interior of the precipitate phase. Another possible scenario can be the formation of percolating networks of C vacancies which act as energetically favored pathways for hydrogen to diffuse from the interface to the bulk of the carbide, as observed for TiC [10]. We can further define the hydrogen–vacancy complex formation energy as the sum of the C vacancy formation energy and the H trapping energy within this vacancy. The comparison for C vacancies at different positions (near the interface) shows that this energy is lowest for vacancies at C-terminated interfaces and, hence, such a scenario is, energetically, the most favorable.

3.4. H Enhanced Decohesion of Interfaces

It is a decisive follow-up question whether the trapped H in the interface causes decohesion along the interface, which might serve as a mechanism of hydrogen embrittlement. To understand this, we need to answer the question whether the presence of H in the interface enables or suppresses the separation of the interface into respective free surfaces. To do so, the adsorption energy of H at the interface was compared to the energy of H at the free surfaces of the carbide and the austenitic Fe matrix after possible crack formation. The corresponding energy of separation is given by the expression:

$$E_{sep} = [E_{el}(\text{int}) - E_{el}(\kappa\text{-surf}) - E_{el}(\text{Fe-surf})]$$
$$+ [c_H(\text{int}) E_{trap}(\text{int}) - c_H(\kappa) E_{ads}(\kappa\text{-surf}) - c_H(\text{Fe}) E_{ads}(\text{Fe-surf})] \qquad (4)$$

where $E_{el}(\text{int})$ is the total energy of the supercell containing the interface of κ-carbide and fcc-Fe without H. $E_{el}(\kappa\text{-surf})$ and $E_{el}(\text{Fe-surf})$ are the total energies of supercells containing κ-carbide and fcc-Fe, respectively, with a surface adjacent to a vacuum layer. c_H is the concentration of H at the planar defects, which depends on its chemical potential μ_H. $E_{trap}(\text{int})$ and $E_{ads}(X\text{-surf})$ are the corresponding trapping and adsorption energies of hydrogen at the interface or the surface ($X = \kappa$ or Fe) given by:

$$E_{trap}(\text{int}) = E_{el}(\text{H@int}) - E_{el}(\text{int}) - \mu_H \text{ and}$$
$$E_{ads}(X\text{-surf}) = E_{el}(\text{H@X-surf}) - E_{el}(X\text{-surf}) - \mu_H \qquad (5)$$

The tendency of trapped H to cause decohesion at the interface is indicated by the "embrittlement term" as given by the second term in Equation (4), namely:

$$\varepsilon = c_H(\text{int})E_{trap}(\text{int}) - c_H(\kappa)E_{ads}(\kappa\text{-surf}) - c_H(\text{Fe})E_{ads}(\text{Fe-surf}). \qquad (6)$$

A positive value of ε implies that it is energetically favorable to have hydrogen on free surfaces rather than at the interface, which provides a driving force for the formation of free surfaces. On the other hand, a negative value indicates that H prefers to stay at the interface, thereby stabilizing it.

The equilibrium concentration of hydrogen at the interface is given by its trapping energy, i.e., $c_H(\text{int}) = 1/(\exp(\beta E_{trap}(\text{int})) + 1)$ with $\beta = 1/(k_B T)$, and depends via Equation (5) on the chemical potential μ_H (Figure 9). We compare an interface without C vacancies (solid line) with an interface that contains a single C vacancy (dashed line). In the first case, a maximum concentration of 1 H atom per 2×2 interface area is considered (corresponding to $c_H = 1$), to avoid strong H–H interactions. In the second case, only the filling of the single vacancy per 2×2 interface area with H is considered. Due to the lower trapping energy, the occupation of the vacancy sets in at much lower chemical potentials than for the perfect interface. The concentration of H at the free surfaces depends on the kinetics of the separation process. If it is very fast, then the particle conservation $c_H(\kappa) + c_H(\text{Fe}) = c_H(\text{int})$ can be assumed, corresponding to the canonical limit. If the diffusibility of H is faster than the separation, then an equilibrium H concentration $c_H(X) = 1/(\exp(\beta E_{ads}(X\text{-surf})) + 1)$ can also be used for the surfaces, corresponding to the grand-canonical limit. Our result points in this limit towards a strong driving force for H to segregate to a free κ-carbide surface with respect to a position at the interface (Figure 9).

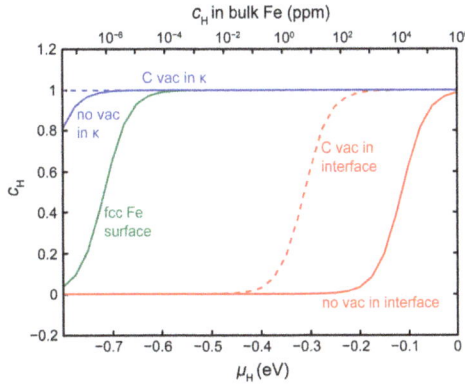

Figure 9. Equilibrium concentration of H at the C-terminated interface, free κ-carbide (C-terminated) and fcc-Fe surfaces as a function of the chemical potential of hydrogen. A perfect κ-carbide (solid lines) is compared with a κ-carbide that contains a C vacancy (dashed lines).

Based on these concentrations, the embrittlement term was determined for both limits according to Equation (6) and plotted in Figure 10. Looking first at the results for the perfect interface in the grand-canonical (GC) limit, we observe large positive values for all considered chemical potentials of H. This is because the free surfaces of κ-carbide and fcc-Fe are substantially more favorable for H than the interface. A similar effect has been observed for several other precipitate phases [41]. We note, however, that this is a purely thermodynamic argument. It ignores the fact that the H concentration at the perfect interface is negligibly small for $\mu_H < -0.3$ eV and that kinetics would not provide H significantly fast in order to stabilize the forming free surfaces. Therefore, we have plotted the canonical (C) limit in addition, for which the embrittlement term is almost vanishing for these chemical potentials. The processes in reality will occur between these two limits.

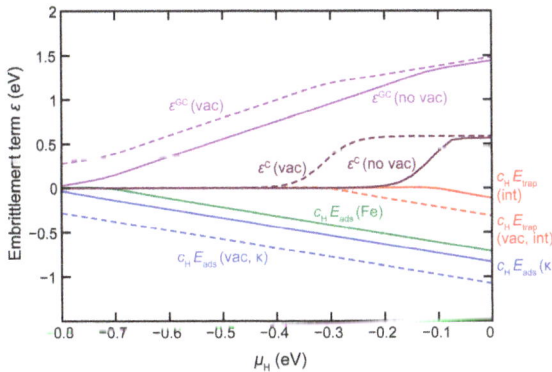

Figure 10. Embrittlement term ε and the different energy contributions in Equation (6) as a function of μ_H. A perfect κ-carbide (solid lines) is compared with a κ-carbide that contains a C vacancy (dashed lines). Furthermore, the grand-canonical (GC) and the canonical (C) limit are compared (see text for details).

In order to come closer to reality, it is also important to consider the existence of C vacancies in the interface. Due to the substantial trapping effect of these defects, these sites will be quickly filled with H for chemical potentials $\mu_H > -0.5$ eV. This affects the first term of Equation (6), indicated by the red dashed line in Figure 10. At the same time, the H incorporation at the forming κ-carbide surface, which contains this vacancy as well, also becomes even more favorable (blue dashed line in Figure 10). Depending on the H kinetics, i.e., the grand-canonical vs. the canonical limit, the combined embrittlement term ε increases further (purple and maroon dashed lines in Figure 10). Since in particular the canonical line shifts substantially upwards, we can conclude that the presence of these lattice imperfections is decisive for a destabilization of the interface if H comes into play.

4. Conclusions

The potential of κ-carbides to trap diffusive hydrogen was studied using density-functional theory. The hydrogen analog of the κ-carbide, the κ-hydride, was shown to be a theoretically stable compound. Its formation, however, is kinetically disadvantageous when compared to the carbide.

Substituting Mn for Fe was clearly shown to be preferable to the trapping energy, since the chemical potential of hydrogen decreases linearly with the number of manganese atoms in the octahedron surrounding the hydrogen atom. This effect is independent of the overall carbon concentration although directly neighboring carbon atoms seem to reduce the Mn–H attraction. As a result, the trapping capability of κ-carbides is proportional to the manganese content of the carbide.

The interaction of hydrogen with different metal matrices, including the $L1_2$-Fe_3Al matrix, was investigated, and it was observed that the κ-matrix is a favorable environment for H as a result of both its chemical environment, owing to attractive Al–H interactions in κ-like arrangements, and the increased lattice parameter. The interaction between carbon and hydrogen as well as between different hydrogen atoms within the κ-matrix was examined. Both the C–H and the H–H interaction proved to be repulsive, with the H–H interaction showing a stronger repulsion in the short range while C–H repulsion is stronger in the long range. Additionally, directly neighboring C and H atoms suppress the long-range interaction.

For κ-carbides with a carbon content of about 35–75% with respect to the stoichiometric compound, the repulsion between the interstitials is compensated by the increase in trapping capability due to an increase of the lattice parameter. The average trapping energy of a fully filled κ-carbide is, thus, independent from the carbon loading of the carbide.

We conclude that the trapping efficiency of diffusive hydrogen by κ-carbides depends on its C, as well as its Mn, content. While the C content determines the number of vacant κ positions and thereby the capacity of the phase to trap hydrogen, the manganese content determines the overall trapping efficiency, i.e., the average gain in energy if an H atom is removed from the matrix and stored in the carbide precipitate. Efficient strategies to control diffusive H in steel materials should thus consider lowering the carbon contents of κ-carbide precipitates while raising their manganese contents.

Finally, we demonstrate that hydrogen located in the interface between carbides and the matrix material has a high solubility only in the presence of vacancies near the interface, thereby making the carbide/austenite interface an efficient trapping center for hydrogen. Our DFT calculations further suggest a possible hydrogen-induced separation of the interface into free surfaces of κ-carbide and fcc-Fe, thereby leading to embrittlement caused by a hydrogen-enhanced decohesion mechanism. Said embrittlement can be expressed via a quantitative model to assess the dependence of the interface decohesion upon hydrogen concentration.

Acknowledgments: This work is part of the Collaborative Research Center 761 "Steel ab initio" funded by the Deutsche Forschungsgemeinschaft (German Research Foundation, DFG). We gratefully acknowledge their financial support. Furthermore, we gratefully thank Salzgitter Mannesmann Forschung GmbH (SZMF) for their cooperation and funding of the project. We would also like to thank the JARA-HPC supercomputing center (ITC) of RWTH Aachen University for providing computational resources within grant JARA0057.

Author Contributions: Tobias A. Timmerscheidt, Dimitri Bogdanovski, and Jörg von Appen conceived and performed all simulations concerned with C–H/H–H interactions and the influence of Mn in bulk κ-carbide; Poulumi Dey and Tilmann Hickel conceived and performed all simulations focusing on the interface and conceptualized the embrittlement term; Tobias A. Timmerscheidt, Poulumi Dey, Dimitri Bogdanovski, and Jörg von Appen performed the data analysis and wrote the initial draft of the manuscript. Jörg Neugebauer and Richard Dronskowski initiated and supervised the project. All authors contributed equally to the interpretation of the data and the writing of the final version of the paper.

Conflicts of Interest: The authors declare no conflict of interest.

References

1. Gutierrez-Urrutia, I.; Raabe, D. Influence of Al content and precipitation state on the mechanical behavior of austenitic high-Mn low-density steels. *Scr. Mater.* **2013**, *68*, 343–347. [CrossRef]

2. Park, K.-T.; Kim, G.; Kim, S.K.; Lee, S.W.; Hwang, S.W.; Lee, C.S. On the transitions of deformation modes of fully austenitic steels at room temperature. *Met. Mater. Int.* **2010**, *16*, 1–6. [CrossRef]

3. Gutierrez-Urrutia, I.; Raabe, D. High strength and ductile low density austenitic FeMnAlC steels: Simplex and alloys strengthened by nanoscale ordered carbides. *Mater. Sci. Technol.* **2014**, *30*, 1099–1104. [CrossRef]

4. Zambrano, O.A.; Valdés, J.; Aguilar, Y.; Coronado, J.J.; Rodríguez, S.A.; Logé, R.E. Hot deformation of a Fe-Mn-Al-C steel susceptible of κ-carbide precipitation. *Mater. Sci. Eng. A* **2017**, *689*, 269–285. [CrossRef]

5. Haase, C.; Zehnder, C.; Ingendahl, T.; Bikar, A.; Tang, F.; Hallstedt, B.; Hu, W.; Bleck, W.; Molodov, D.A. On the deformation behavior of κ-carbide-free and κ-carbide-containing high-Mn light-weight steel. *Acta Mater.* **2017**, *122*, 332–343. [CrossRef]

6. Hirth, J.P. Effects of hydrogen on the properties of iron and steel. *MTA* **1980**, *11*, 861–890. [CrossRef]

7. Hojo, T.; Sugimoto, K.-I.; Mukai, Y.; Ikeda, S. Effects of Aluminum on Delayed Fracture Properties of Ultra High Strength Low Alloy TRIP-aided Steels. *ISIJ Int.* **2008**, *48*, 824–829. [CrossRef]

8. Song, S.W.; Kwon, Y.J.; Lee, T.; Lee, C.S. Effect of Al addition on low-cycle fatigue properties of hydrogen-charged high-Mn TWIP steels. *Mater. Sci. Eng. A* **2016**, *677*, 421–430. [CrossRef]

9. Takahashi, J.; Kawakami, K.; Kobayashi, Y.; Tarui, T. The first direct observation of hydrogen trapping sites in TiC precipitation-hardening steel through atom probe tomography. *Scr. Mater.* **2010**, *63*, 261–264. [CrossRef]

10. Di Stefano, D.; Nazarov, R.; Hickel, T.; Neugebauer, J.; Mrovec, M.; Elsässer, C. First-principles investigation of hydrogen interaction with TiC precipitates in α-Fe. *Phys. Rev. B* **2016**, *93*, 184108. [CrossRef]

11. Dierkes, H.; van Leusen, J.; Bogdanovski, D.; Dronskowski, R. Synthesis, Crystal Structure, Magnetic Properties, and Stability of the Manganese-Rich "Mn_3AlC" κ Phase. *Inorg. Chem.* **2017**, *56*, 1045–1048. [CrossRef] [PubMed]

12. Andryushchenko, V.A.; Gavrilyuk, V.G.; Nadutov, V.M. Atomic and magnetic ordering in k-phase of Fe-Al-C alloys. *Phys. Met. Metallogr.* **1985**, *60*, 50–55.

13. Seol, J.-B.; Raabe, D.; Choi, P.; Park, H.S.; Kwak, J.H.; Park, C.G. Direct evidence for the formation of ordered carbides in a ferrite based low-density Fe–Mn–Al–C alloy studied by transmission electron microscopy and atom probe tomography. *Scr. Mater.* **2013**, *68*, 348–353. [CrossRef]

14. Yao, M.J.; Dey, P.; Seol, J -B.; Choi, P.; Herbig, M.; Marceau, R.K.W.; Hickel, T.; Neugebauer, J.; Raabe, D. Combined atom probe tomography and density functional theory investigation of the Al off-stoichiometry of kappa-carbides in an austenitic Fe-Mn-Al-C low density steel. *Acta Mater.* **2016**, *106*, 229–238.

15. Dey, P.; Nazarov, R.; Dutta, B.; Yao, M.; Herbig, M.; Friák, M.; Hickel, T.; Raabe, D.; Neugebauer, J. Ab initio explanation of disorder and off-stoichiometry in Fe-Mn-Al-C κ carbides. *Phys. Rev. B* **2017**, *95*, 104108. [CrossRef]

16. Palm, M.; Inden, G. Experimental determination of phase equilibria in the Fe-Al-C system. *Intermetallics* **1995**, *3*, 443–454.

17. Desai, S.K.; Neeraj, T.; Gordon, P.A. Atomistic mechanism of hydrogen trapping in bcc Fe-Y solid solution: A first principles study. *Acta Mater.* **2010**, *58*, 5363–5369. [CrossRef]

18. Psiachos, D.; Hammerschmidt, T.; Drautz, R. Ab initio study of the interaction of H with substitutional solute atoms in α-Fe: Trends across the transition-metal series. *Comput. Mater. Sci.* **2012**, *65*, 235–238. [CrossRef]

19. Nazarov, R.; Hickel, T.; Neugebauer, J. Ab initio study of H-vacancy interactions in fcc metals: Implications for the formation of superabundant vacancies. *Phys. Rev. B* **2014**, *89*, 144108. [CrossRef]

20. Du, Y.A.; Ismer, L.; Rogal, J.; Hickel, T.; Neugebauer, J.; Drautz, R. First-principles study on the interaction of H interstitials with grain boundaries in α- and γ-Fe. *Phys. Rev. B* **2011**, *84*, 144121. [CrossRef]

21. Baligidad, R.G.; Prakash, U.; Radhakrishna, A.; Ramakrishna Rao, V. Effect of carbides on embrittlement of Fe3Al based intermetallic alloys. *Scr. Mater.* **1997**, *36*, 667–671. [CrossRef]

22. Sen, M.; Balasubramaniam, R. Hydrogen trapping at carbide-matrix interfaces in Fe₃AlC intermetallics. *Scr. Mater.* **2001**, *44*, 619–623. [CrossRef]

23. Koyama, M.; Springer, H.; Merzlikin, S.V.; Tsuzaki, K.; Akiyama, E.; Raabe, D. Hydrogen embrittlement associated with strain localization in a precipitation-hardened Fe–Mn–Al–C light weight austenitic steel. *Int. J. Hydrogen Energy* **2014**, *39*, 4634–4646. [CrossRef]

24. Kresse, G.; Hafner, J. Ab initio molecular dynamics for liquid metals. *Phys. Rev. B* **1993**, *47*, 558–561. [CrossRef]

25. Kresse, G.; Hafner, J. Ab initio molecular dynamics simulation of the liquid-metal–amorphous-semiconductor transition in germanium. *Phys. Rev. B* **1994**, *49*, 14251–14269. [CrossRef]

26. Kresse, G.; Furthmüller, J. Efficiency of ab-initio total energy calculations for metals and semiconductors using a plane-wave basis set. *Comp. Mater. Sci.* **1996**, *6*, 15–50. [CrossRef]

27. Kresse, G.; Furthmüller, J. Efficient iterative schemes for ab initio total-energy calculations using a plane-wave basis set. *Phys. Rev. B* **1996**, *54*, 11169–11185. [CrossRef]

28. Blöchl, P.E. Projector-augmented wave method. *Phys. Rev. B* **1994**, *50*, 17953–17979. [CrossRef]

29. Kresse, G.; Joubert, D. From ultrasoft pseudopotentials to the projector augmented-wave method. *Phys. Rev. B* **1999**, *59*, 1758–1774. [CrossRef]

30. Perdew, J.P.; Burke, K.; Ernzerhof, M. Generalized Gradient Approximation Made Simple. *Phys. Rev. Lett.* **1997**, *77*, 3865–3868. [CrossRef] [PubMed]

31. Monkhorst, H.J.; Pack, J.D. Special points for Brillouin-zone integrations. *Phys. Rev. B* **1976**, *13*, 5188–5192. [CrossRef]

32. Methfessel, M.; Paxton, A.T. High-precision sampling for Brillouin-zone integration in metals. *Phys. Rev. B* **1989**, *40*, 3616–3621. [CrossRef]

33. Ekholm, M.; Abrikosov, I.A. Structural and magnetic ground-state properties of γ-FeMn alloys from ab initio calculations. *Phys. Rev. B* **2011**, *84*, 104423. [CrossRef]

34. Lee, K.; Yuan, M.; Wilcox, J. Understanding Deviations in Hydrogen Solubility Predictions in Transition Metals through First-Principles Calculations. *J. Phys. Chem. C* **2015**, *119*, 19642–19653. [CrossRef]

35. Kellou, A.; Raulot, J.M.; Grosdidier, T. Structural and thermal properties of Fe₃Al, Fe₃AlC and hypothetical Fe₃AlX (X = H, B, N, O) compounds: Ab initio and quasi-harmonic Debye modelling. *Intermetallics* **2010**, *18*, 1293–1296. [CrossRef]

36. Connétable, D.; Maugis, P. First principle calculations of the k-Fe₃AlC perovskite and iron/aluminium intermetallics. *Intermetallics* **2008**, *16*, 345–352. [CrossRef]

37. Noh, J.Y.; Kim, H. Ab initio calculations on the effect of Mn substitution in the κ-carbide Fe₃AlC. *J. Korean Phys. Soc.* **2013**, *62*, 481–485. [CrossRef]

38. von Appen, J.; Dronskowski, R.; Chakrabarty, A.; Hickel, T.; Spatschek, R.; Neugebauer, J. Impact of Mn on the solution enthalpy of hydrogen in austenitic Fe-Mn alloys: A first-principles study. *J. Comp. Chem.* **2014**, *35*, 2239–2244. [CrossRef] [PubMed]

39. Song, W.; Zhang, W.; von Appen, J.; Dronskowski, R.; Bleck, W. κ-phase Formation in Fe–Mn–Al–C Austenitic Steels. *Steel Res. Int.* **2015**, *86*, 1161–1169. [CrossRef]

40. Timmerscheidt, T.A.; Dronskowski, R. An Ab Initio Study of Carbon-Induced Ordering in Austenitic Fe-Mn-Al-C Alloys. *Steel Res. Int.* **2017**, *88*, 1600292. [CrossRef]

41. Milella, P.P. *Fatigue and Corrosion in Metals*, 1st ed.; Springer: Milan, Italy, 2013.

metals

MDPI

Article

Transition of the Interface between Iron and Carbide Precipitate From Coherent to Semi-Coherent

Hideaki Sawada [1],*, Shunsuke Taniguchi [1], Kazuto Kawakami [2] and Taisuke Ozaki [3]

[1]   Advanced Technology Research Laboratories, Nippon Steel & Sumitomo Metal Corporation, 1-8 Fuso-Cho, Amagasaki, Hyogo 660-0891, Japan; taniguchi.s8k.shunsuke@jp.nssmc.com
[2]   Nippon Steel & Sumikin Technology Co. Ltd., 20-1 Shintomi, Futtsu, Chiba 293-0011, Japan; kawakami-kazuto@nsst.jp
[3]   The Institute of Solid State Physics, The University of Tokyo, 5-1-5 Kashiwanoha, Kashiwa, Chiba 277-8581, Japan; t-ozaki@issp.u-tokyo.ac.jp
*   Correspondence: sawada.x4d.hideaki@jp.nssmc.com; Tel.: +81-6-6489-5960

Received: 5 May 2017; Accepted: 14 July 2017; Published: 19 July 2017

Abstract: There are some precipitates that undergo transition from a coherent to semi-coherent state during growth. An example of such a precipitate in steel is carbide with a NaCl-type structure, such as TiC and NbC. The interface energy between carbide precipitate and iron is obtained via large-scale first-principles electronic structure calculation. The strain energy is estimated by structure optimization of the iron matrix with virtual carbide precipitate using the empirical potential. The transition of the interface from a coherent to semi-coherent state was examined by comparing the interface and strain energies between the coherent and semi-coherent interfaces. The sizes where both the precipitates undergo this transition are smaller than those of the interfaces with minimum misfit. The estimated transition diameter of TiC is in agreement with the experimentally obtained value.

Keywords: first-principles calculation; interface; iron; precipitate

---

1. Introduction

Strengthening of steel is mainly performed through four mechanisms [1]. Increasing the strength of steels by using a solid solution of alloying elements and impurity elements is known as solid solution strengthening. Another strengthening mechanism is dislocation strengthening. In this mechanism, the yield strength is proportional to the square root of the dislocation density. Grain refinement is another effective strengthening mechanism. In this process, yield strength is inversely proportional to the square root of grain size. Precipitates are obstacles to dislocation movement, resulting in strengthening of steel, known as precipitation strengthening. Among these four mechanisms, the most effective mechanism is precipitation strengthening. The effect of precipitation strengthening is expected to be 3000 MPa for the Orowan mechanism, when the volume fraction and diameter of precipitates are assumed to be 10% and 10 nm, respectively. The strengthening effect of this process is much larger than those of the other three mechanisms. Precipitates in steel contribute to not only strengthening but also reduction of grain size, because grain growth is inhibited by pinning of grain boundaries during thermal treatment [2]. This is an important technique to inhibit softening of the heat-affected zone during welding [3]. Other important roles of precipitates include improvements of hole expandability [4] and the hydrogen trap to prohibit the hydrogen embrittlement [5,6].

Precipitates usually have lattice constants that are dissimilar to those of the matrix, i.e., iron. Elastic strain energy due to precipitates is proportional to the cube of the radius of the precipitate. Therefore, the growth of precipitates substantially increases the elastic strain energy. Dislocations due to the semi-coherent interface between precipitate and matrix can reduce the strain around the precipitate. In steel, some precipitates are nucleated with a coherent interface between the precipitate and iron. Such a coherent precipitate naturally transforms to a semi-coherent precipitate during growth, when the coherent precipitate has higher interface and strain energies than the semi-coherent precipitate [7]. One such precipitate has a NaCl-type structure, such as TiC, NbC, and VC. These precipitates are considered to play important roles in the strength and structure of steel. Semi-coherent precipitates can provide a very different performance in comparison to coherent precipitates, for example, in terms of strength and hydrogen trap. This is because it is assumed that the behavior of dislocations and hydrogen varies according to the strain field around the precipitate. In other words, the performance of steel can be improved by appropriate usage of coherent and semi-coherent precipitates. Therefore, it is important to know the size at which the coherent precipitate transforms to a semi-coherent precipitate. These carbides with nanometer size can be densely distributed in steels. An interface relationship of {001}$_{precipitate}$//{001}$_{iron}$ and <100>$_{precipitate}$//<110>$_{iron}$ is reported as a Baker–Nutting orientation relationship for carbide precipitates with an NaCl-type structure [8]. One such precipitate, TiC, is considered to exist as a coherent precipitate up to a size of 3 nm [9], and the precipitate transforms to a semi-coherent precipitate due to growth.

The precipitate TiC has a plate-like shape with the abovementioned interface relationship between the precipitate and the iron matrix. In other words, the precipitate grows mainly in the broad plane, namely, the (100) plane of the precipitate. The lateral plane of the precipitate is expected to be semi-coherent at the early stage of precipitation and have higher interface energy than the broad plane. Therefore, the growth of the lateral plane is not dominant. In this study, we focused on the dominant growth plane of the precipitate, namely the broad plane. The growth of the lateral plane and the anisotropy of the growth will be the subject of our future work.

Many first-principles calculations have been performed for the interfaces between body-centered cubic (bcc) iron and carbide or nitride precipitates with an NaCl-type structure [10–20], and the bonding nature between these interfaces has been discussed in the literature. Hartford investigated the chemical bond across the interface and interface energies for the coherent interface between VN and bcc iron [10]. VN in bcc iron was observed as a plate-like precipitate with a diameter between 10 and 20 nm. The treatment as a coherent precipitate is quite reasonable because the lattice misfit between bcc iron and VN is less than 2%. The bond between iron and VN was clarified to be a strong $pd\sigma$ bond and a weak $dd\sigma$ bond. Lee et al. investigated the bond strength between iron and a precipitate by separating the interface between bcc iron and TiC [11]. Arya et al. performed similar calculations in order to apply steel to more reactive and corrosive environments by TiC-coating the iron surface [12]. All of their investigations concluded that the bond between Fe and C atoms strongly depends on the short-range interaction due to the $pd\sigma$ bond. In contrast, the bond among Fe atoms is stronger than that between Fe and C atoms in terms of long-range interaction because of the difference between the metallic bond among Fe atoms and the covalent bonds between Fe and C atoms.

The coherent interface energies between iron and a precipitate were estimated by Hartford [10] and Jung et al. [13–16]. Johansson et al. developed a method for determining the interface energy using a few points of γ surface [18–20]. They showed that the strain energy should be relieved by forming a semi-coherent precipitate for thicker precipitates, since the elastic energy caused by the coherent strain increases with the thickness of the precipitate. In their approach, the interface energy was estimated using the extended Peierls–Nabarro framework, in which the chemical interaction energy by the first-principles calculation is combined with the elastic energy by the continuum description. They explained that the elastic contribution dominates the interface energy. By calculating the interface energies of many 3d, 4d, and 5d transition metal carbides and nitrides, it was revealed that the interface energies decrease with increase in the number of electrons of the outer-most d orbital [19]. This phenomenon was interpreted as the stronger chemical bond between the Fe atom and transition metal atom in carbides or nitrides for the transition metal atom with more d electrons.

The computational technique by Fors et al. obtains the semi-coherent interface energy efficiently by using the γ surface energy and elastic strain energy using the Peierls–Nabarro framework. However, there is an ambiguity in the accuracy of the energy value because the atomic configuration around the misfit dislocation is not reproduced accurately. Furthermore, it is not easy for their technique to accurately obtain the segregation properties of impurity near the interface. Previously, we tried to obtain the interface energy of semi-coherent interfaces between NbC and bcc iron via large-scale direct first-principles calculation at the semi-coherent interface [21]. We estimated the transition size of NbC from coherent to semi-coherent precipitate based on the interface and strain energies obtained via the large-scale first-principles calculation and the classical molecular dynamics simulation, respectively. However, the dependence of the precipitate on the size at which the coherent to semi-coherent interface transition occurs has not been clarified to date. Therefore, herein, we applied a method to estimate the transition size of TiC and compared it with the previous results of NbC as well as the interface and strain energies, the electronic structure, and the atomic configuration.

2. Results and Discussion

2.1. Interface Energy

The conditions of calculating coherent interface energy, which include the shape of the unit cell shown in Figure 1, are followed according to the previous study [21]. Two cases of coherent interfaces are examined, as performed in the previous study. In these cases, the C atom is directly connected with the Fe atom across the interface and the Ti atom is connected with the Fe atom. The cases are hereafter referred to as the interfaces of Fe–C–Ti and Fe–Ti–C, respectively. The interface energies are calculated by subtracting the sum of the total energies of all the individual phases from the total energy of the system including the interface. The lattice parameter of the reference state of iron is chosen as follows in order to eliminate the contribution of strain energy in the estimation of the interface energy. The lattice parameter c_0 is optimized to minimize the total energy with the a_0 value fixed to the length of a unit cell of TiC. This procedure is conducted for both the coherent and semi-coherent interfaces. The lattice parameter a_0 of iron for a coherent interface is $a_0(\text{TiC})$, while that for semi-coherent interface is $(n/m)a_0(\text{TiC})$. In Figure 2, the calculated interface energies are shown with the previously obtained interface energies between the NbC and bcc iron. In the previous study, two coherent interfaces were treated. One is an interface where a C atom is connected with an Fe atom across the Fe–NbC interface; the other is one where a Nb atom is connected with an Fe atom across the Fe–NbC interface. These two interfaces are referred as Fe–C–Nb and Fe–Nb–C, respectively, in this paper. The horizontal and vertical axes denote the interface distance d and interface energy, respectively. The interface distance is defined as the optimized length of c_0 of $\text{Fe}_7\text{Ti}_7\text{C}_7$ by subtracting the optimized lengths of c_0 of Fe and TiC with a_0 of TiC as follows:

$$d = \{c_0(\text{Fe}_7\text{Ti}_7\text{C}_7) - c_0(\text{Fe}_6) - c_0(\text{Ti}_6\text{C}_6)\}/2. \tag{1}$$

The division operation in this expression is performed because of the inclusion of two interfaces in the unit cell. The interface of Fe–C–Ti is more stable than that of Fe–Ti–C. The same applies for the interface between NbC and bcc iron, which is explained as the energy of the atom at the interface of Fe–C–Nb is lower than that of Fe–Nb–C in the previous study [21]. The local density of states (DOSs) of the C atom of the interface between TiC and bcc iron are plotted in Figure 3. The phenomenon that occurs in the interface between TiC and bcc iron is similar to that between the NbC and bcc iron. The local DOS of the C atom at the interface is shifted to a higher energy relative to that at the middle of the TiC region in the interface of Fe–Ti–C, while the energy region of local DOS of the C atom at the interface is close to that at the middle of the TiC region in the interface of Fe–C–Ti. The mechanism of the energy shift of the local DOS of the C atom at the interface of Fe–Ti–C can be related to the number of surrounding atoms of the C atom. In TiC with an NaCl-type structure, the C atom is surrounded by six Ti atoms. However, the C atom at the interface of Fe–Ti–C has only five surrounding atoms, i.e., the C atom has a dangling bond that destabilizes the C atom. Conversely, the C atom at the interface has six surrounding atoms at the interface of Fe–C–Ti. The local DOS of the Ti atom at the interface of Fe–Ti–C also shows the energy shift in Figure 3d, but the shift is more clearly seen in the DOS of the C atom. Thus, the electronic effect for the interface bonding is dominated by the electrons around C rather than Ti. The states of the C atom in TiC are mainly located from −6 to 0 eV in Figure 3a,b, while those in NbC are from −7 to −2 eV, as shown in the previous paper [21]. In other words, the energy band of TiC is wider than that of NbC. This is due to the difference of lattice constants between TiC and NbC. The smaller lattice constant of TiC gives a wider energy band of the C atom in comparison with NbC. This spread electron of C atom in TiC reduces the difference of the local DOS of the C atom between the interfaces of Fe–C–Ti and Fe–Ti–C compared to that between the interfaces of Fe–C–Nb and Fe–Nb–C, because the energy shift originates from the local covalent bond character of the C atom at the interface.

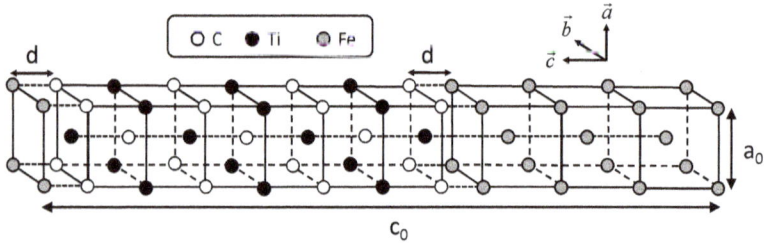

Figure 1. Arrangement of atoms in the coherent interface with 14 layers along the c-axis.

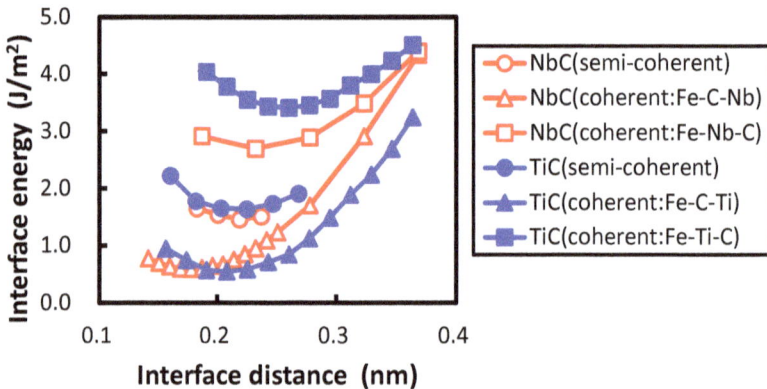

Figure 2. The interface distance dependence on the interface energies of the coherent and semi-coherent interfaces between the TiC and bcc iron and those between the NbC and body-centered cubic (bcc) iron.

The minimum of interface energy of Fe–Nb–C is much lower than that of Fe–Ti–C. In the transition metal carbide the transition metal d states hybridize with the $2p$ states of carbon, and the bonding and anti-bonding states are separated distinctly (Figure 3). For the interfaces of Fe–Ti–C and Fe–Nb–C, the bond across the interface is attributed to the metallic bond between the transition metal d states and Fe d states. In such a case, the increase in filling of the transition metal d states increases the strength of the interface bond [19]. Basically, Nb has one more electron than Ti in the outer-most d orbital. Furthermore, the number of electrons in the Nb atom at the interface increases compared with that at the middle of the NbC region by approximately 0.1 due to the electron donation from the Fe atom, while the number of electrons of Ti at the interface is almost the same as that at the middle of the TiC region. The deeper electronic states of Nb in comparison to Ti cause this phenomenon. Therefore, the interface bond of Fe–Nb–C should be stronger than that of Fe–Ti–C, and the interface of Fe–Nb–C has lower interface energy than that of Fe–Ti–C. In contrast, the minimum of interface energy of Fe–C–Ti does not differ from that of Fe–C–Nb. The bond across the interface should be derived from the strong covalent bond between the Fe and C atoms for both interfaces. Therefore, there is no large deviation in the interface energy between Fe–C–Nb and Fe–C–Ti. In addition, as shown in Figure 2, the interface separation in the interface of Fe–C–Ti is shorter than that of Fe–Ti–C, and the interface separation in the interface of Fe–C–Nb is shorter than that of Fe–Nb–C. The shorter interface separation of Fe–C–Ti in comparison with that of Fe–Ti–C originates from the strong covalent bond between the Fe and C atoms across the interface of Fe–C–Ti.

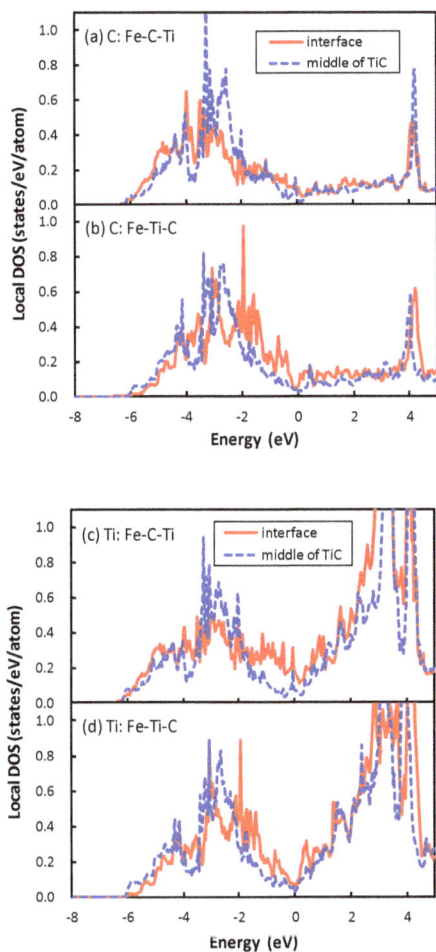

Figure 3. Local density of states (DOS) of the C atoms (**a**,**b**) and Ti atoms (**c**,**d**) located at the interface and the middle of TiC. For the atomic arrangement at the interface, (**a**,**c**) correspond to the case that the C atom is directly connected with the Fe atom across the interface; and (**b**,**d**) correspond to the case that the Ti atom is connected with the Fe atom. The solid red lines and blue dashed lines denote the local DOS at the interface and the middle of TiC, respectively. The Fermi level is set as 0.

In general, m units of bcc iron face n units of TiC at the interface. The interface with the same m and n is coherent, while that with different m and n is semi-coherent. In many cases, m and n cannot be expressed by a ratio of integers, i.e., an incoherent interface. As aforementioned, precipitate nanometer sizes of TiC in iron have a flat interface represented by the Baker–Nutting orientation relationship and reveal the semi-coherent interface [9]. It is possible to find the minimum of misfit Δa by searching m and n with lattice constant of iron $a_0(\text{Fe})$ and that of TiC $a_0(\text{TiC})$. $a_0(\text{TiC})$ is $1/sqrt(2)$ of the conventional lattice constant.

$$\Delta a = |m \times a_0(\text{Fe}) - n \times a_0(\text{TiC})|. \tag{2}$$

For the interface between bcc iron and TiC, $m = 15$ and $n = 14$ afford a minimum misfit Δa. Large-scale first-principles calculations are performed for $m{:}n = 15{:}14, 14{:}13, 13{:}12, 12{:}11, 11{:}10$. The interface energies calculated for $m = 15$ and $n = 14$ are shown in Figure 2.

The interface distance of the semi-coherent interface is defined as the same analogy of Equation (1). The interface distance of the semi-coherent interface between the TiC and bcc iron is close to that of the coherent interface of Fe–C–Ti. In contrast, the interface distance of the semi-coherent interface between the NbC and bcc iron is closer to that of the coherent interface of Fe–Nb–C. This discrepancy occurs for two reasons. First, the interface distance of the coherent interface between NbC and bcc iron is underestimated compared with that between the TiC and bcc iron, because the lattice parameter of the iron is expanded along *a* axis to the lattice parameter of transition metal carbides. In other words, the unit cell length of a_0 for the coherent interface between the NbC and bcc iron is longer than that between the TiC and bcc iron by 5%. The second cause of the discrepancy comes from a difference in the lattice constant between TiC and NbC. The difference of lattice constants between the TiC and bcc iron is 7.2%, while that between NbC and bcc iron is 12.6%. This large difference of the lattice constants between the TiC and NbC leads to the difference of distance between the misfit dislocations of the semi-coherent interface. The coherent interface distance of Fe–C–Ti is much shorter than that of Fe–Ti–C because of the different bonding character between the covalent strong Fe–C bond and the metallic Fe–Ti bond. There are areas where a Fe atom is connected with a Ti atom across the interface near the misfit dislocation, leading to separation of the interface. The distance between the misfit dislocations of the semi-coherent interface between TiC and bcc iron is approximately 1.67 times longer than that between the NbC and bcc iron. Therefore, the expansion effect of interface by the misfit dislocation does not extend to the whole interface region for TiC.

The optimized atomic positions of the semi-coherent interface are shown in Figure 4. The Fe atoms at the interface are connected with the C atoms across the interface except for the Fe atoms near the misfit dislocation, which leads to the bending of lines of Fe atoms. A Fe atom can create a strong covalent bond with a C atom, while the bond between Fe and Ti atoms is weak, as described for the coherent interface. Therefore, the bonding character among Fe, Ti, and C atoms reflects on the atomic transfers around the semi-coherent interface.

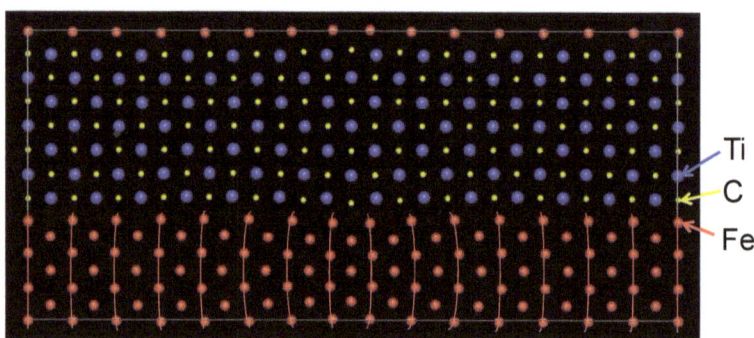

Figure 4. Optimized arrangement of atoms in the semi-coherent interface with $m{:}n = 15{:}14$ in the (110) plane. The Fe atoms are connected along the [001] direction using red lines for visualization.

The semi-coherent interface can be considered as a mixture of the two coherent interfaces; therefore, the interface energies of the semi-coherent interface are influenced by the interface energies of the coherent interfaces. Since the C atoms at the interface are connected with the Fe atoms at the largest part of the semi-coherent interface, the semi-coherent interface energies are mainly affected by the coherent interface energies of Fe–C–Ti or Fe–C–Nb. Based on this coherent interface energies of Fe–C–Ti or Fe–C–Nb, the coherent interface energies of Fe–Ti–C or Fe–Nb–C affect the semi-coherent interface energies. Then, the semi-coherent interface energy between TiC and bcc iron is slightly larger than that between NbC and bcc iron because the coherent interface energy between TiC and bcc iron is larger than that between NbC and bcc iron. This same tendency was also reported by Fors et al. [19].

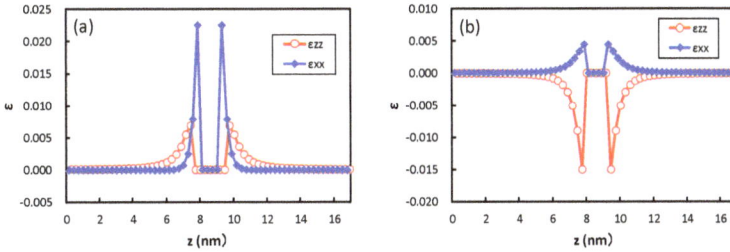

Figure 5. Strain at the center of the broad plane of the precipitate along the direction perpendicular to the broad plane of precipitate with an (**a**) coherent and (**b**) semi-coherent interface. The *x*- and *z*-directions are parallel and perpendicular to the broad plane of the precipitate, respectively.

2.2. Strain Energy

The structure optimization using the empirical potential is performed to obtain the strain around the precipitate in bcc iron. Figure 5 shows the strain obtained at the center of the broad plane of the precipitate along the perpendicular direction to the broad plane of the precipitate. Coherent and semi-coherent precipitates were assumed in Figure 5a,b, respectively. The coherent and semi-coherent precipitates indicate much different features in the strain around the precipitates. The coherent precipitate has a much larger strain, ε_{xx}, than the semi-coherent precipitate. Furthermore, the coherent precipitate gives a positive ε_{zz}, i.e., tensile strain, while the semi-coherent precipitate exhibits compressive strain. These features can be elucidated by the atomic transfers due to precipitation. Figure 6 shows the transfers of atoms around the coherent and semi-coherent precipitate by the structure optimization from the ideal position of Fe atoms in bcc iron due to precipitation. For the coherent precipitate, the atoms at the side of the precipitate move toward [100] or [$\bar{1}$00], because the lattice constant of the precipitate is longer than that of bcc iron. Then, the atoms facing the broad plane of the precipitate also move toward [100] or [$\bar{1}$00]. Thus, atomic density decreases and the atoms move toward the precipitate near the center of broad plane of the precipitate (Figure 6a). This is the origin of tensile strain ε_{zz} for the coherent precipitate. Conversely, atoms at the side of semi-coherent precipitate do not move toward [100] or [$\bar{1}$00]. Then, the decrease in atomic density does not occur near the center of the broad plane of the precipitate, and atoms depart from the precipitate along [00$\bar{1}$] due to the expansion of the precipitate along [001]. In this case, three layers of precipitate are assumed. This precipitate region is constructed using four layers of iron expanded 4% along [001]. Thus, the compressive strain is generated in ε_{zz} for the semi-coherent precipitate. The strain distribution evaluated from atomic arrangement via high angle annular dark-field scanning transmission electron microscopy also reveals compressive strain at the matrix region near the broad plane of precipitate [22]. Therefore, the theoretically estimated strain around the precipitate is qualitatively in good agreement with the experimental observation. The transfers of atoms along [100] for a coherent precipitate are more prominent than those for a semi-coherent precipitate near the broad interface, which leads to larger ε_{xx} for a coherent precipitate than a semi-coherent precipitate.

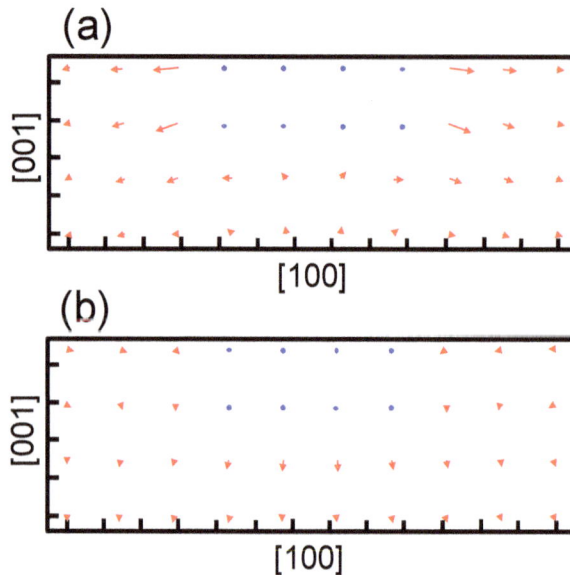

Figure 6. Movement of atoms around the precipitate with an (**a**) coherent and (**b**) semi-coherent interface via structure optimization. The blue points denote atoms in the precipitate region. The red arrows denote movements of atoms in the matrix region by the structural optimization from the ideal position of Fe atoms in bcc iron due to precipitation. The length is elongated by five times from the original for clarity. The tick mark spacing is 0.2 nm for both the horizontal and vertical axes.

Strain energy is estimated by comparing the potential energies of the optimized atoms in the matrix region because of the virtual precipitate region to those with the original atom positions of bcc iron. The obtained strain energy is confirmed to be in good agreement with the continuous model. As mentioned above, we focus on the growth of the broad plane of the precipitate. The interface energy is obtained for the area of the broad plane. The obtained strain energy with the same area of the lateral plane is divided by the area of the broad plane and defined as a function of the area of the broad plane in order to compare it with the interface energy. In Figure 7, the interface and strain energies of NbC and TiC are shown as functions of the diameter of the precipitate. The diameter of the plate-like rectangular parallelepiped shaped precipitate is estimated by assuming the precipitates to be spherical. The thicknesses of the precipitates are 0.54 nm and 0.52 nm for NbC and TiC, respectively. The strain energy strongly depends on the size of the precipitate, in contrast to the small precipitate size dependence of interface energies. The strain energies of coherent precipitates increase proportionally to the diameter of the equivalent sphere of the precipitate because larger strain exists around the precipitate for larger precipitates. This linear dependence of the strain energy on the diameter of the precipitate is consistent with the continuum mechanics. Conversely, for semi-coherent precipitates the strain energies decrease with an increase in the size of precipitate in these regions. In Figure 7, the size of precipitates increases from 1.97 nm ($n = 10$) to 2.51 nm ($n = 14$) for TiC and from 1.22 nm ($n = 5$) to 1.73 nm ($n = 9$) for NbC. The misfits estimated using Equation (2) decrease with increasing size of precipitates in these regions. From this perspective, the misfit and strain energy of semi-coherent interface increases from $n = 15$ to 21 for TiC. In other words, the strain energy of the semi-coherent interface changes in a saw-like shape. The size dependence of strain energies of NbC is much steeper than that of TiC. This can be attributed to the larger lattice misfit between NbC and bcc iron than that between TiC and bcc iron.

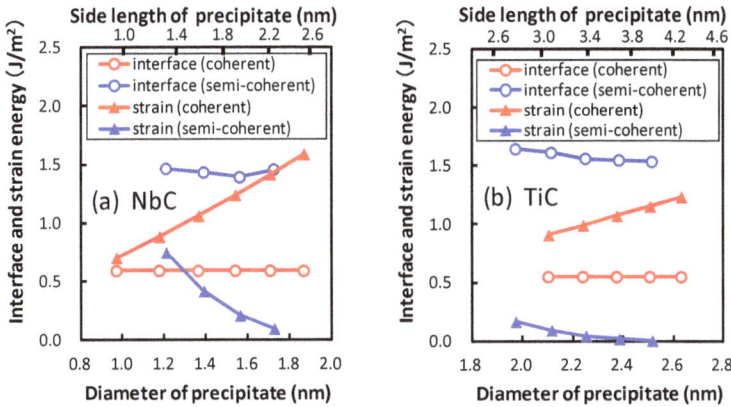

Figure 7. The dependence of diameter of the precipitate on the interface and strain energies for the coherent and semi-coherent precipitates of (**a**) NbC and (**b**) TiC.

2.3. Transition from the Coherent to the Semi-Coherent State

The relation between the summation of the interface and strain energies and the diameter of the precipitate for both the coherent and semi-coherent precipitates is shown in Figure 8. Figure 8 reveals that the coherent precipitates are more stable than the semi-coherent precipitates for smaller size of the precipitates. The opposite size dependence of strain energy between the coherent and semi-coherent precipitates leads to a change from the coherent to the semi-coherent state during the growth of the precipitate at diameters of 1.5 and 2.3 nm for NbC and TiC, respectively. The misfit estimated by Equation (2) has a minimum at diameters of 1.7 and 2.5 nm, respectively. In other words, the sizes where both the precipitates change from coherent to semi-coherent state are smaller than the sizes with minimum misfit in order to release the large strain energies of the coherent precipitates. The transition sizes are examined for thicker precipitates of NbC and TiC with thicknesses of 0.89 and 0.87 nm, respectively. The obtained transition diameters are 1.6 and 2.6 nm for NbC and TiC, respectively.

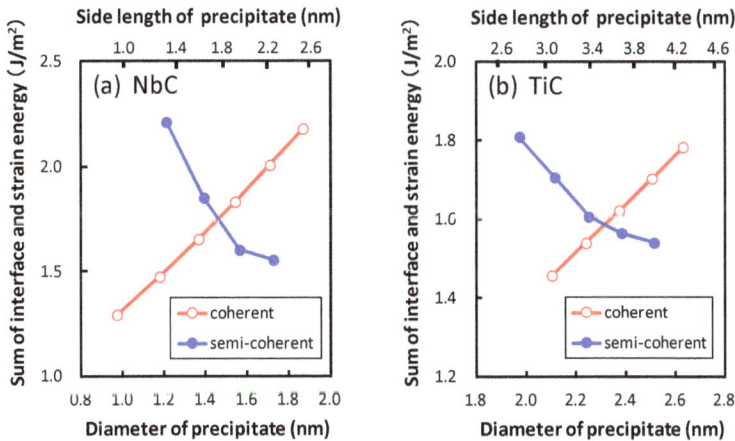

Figure 8. The dependence of diameter of the precipitate on the sum of the interface and strain energies for the coherent and semi-coherent precipitates of (**a**) NbC and (**b**) TiC.

Wei et al. observed interfacial misfit dislocations on plate-like TiC precipitate via high-resolution transmission electron microscopy [9]. They mentioned that one of the semi-coherent precipitates has a broad plane with a length of 6 nm. The diameter of the semi-coherent precipitate is 3–4 nm, if the thickness of the precipitate is assumed to be 0.52 or 0.87 nm. This size is larger than the estimated transition diameter from the coherent to the semi-coherent state. Therefore, the estimated value is consistent with their experimental result.

Kobayashi et al. evaluated the precipitate size dependence of the interaction force on a dislocation due to the precipitate of TiC from the amount of particle strengthening and the particle spacing [23]. They found that the precipitate size dependence of the interaction force due to the precipitate of TiC changes at a diameter of 2.5 nm. The interaction force increases with increasing size of precipitate below the size, but it becomes constant above the size. They speculated that there are several candidates that cause changes in the size dependence of the interaction force. One of them is the transformation from a coherent precipitate to a semi-coherent one. The theoretically predicted transition size from a coherent to semi-coherent precipitate of TiC is close to the precipitate size where the size dependence of the interaction force changes. If the change in the size dependence of the interaction force is attributed to the change from a coherent precipitate to semi-coherent one, the prediction technique of the change from coherent to semi-coherent precipitate gives a fairly accurate value.

3. Computational Details

The interface and strain energies due to the precipitate in the ferritic matrix should be estimated to know the coherency of the interface between the precipitate and matrix. The previous study revealed that the unit cell comprising seven layers each of Fe and NbC in the perpendicular direction of the interface is sufficient to obtain adequate convergence in the interface energy between NbC and bcc iron [21]. Herein, therefore, the same periodic boundary condition is used in the perpendicular direction of the interface. This calculation condition can be justified for the interface between the TiC and bcc iron because the physical properties of TiC are close to those of NbC. For example, the Young's moduli of TiC and NbC are 375 and 330 GPa, respectively, which are 191% and 168% of bcc iron, respectively. Moreover, both TiC and NbC have the same a NaCl-type crystal structure with lattice parameters of 0.433 and 0.447 nm, respectively, and the lattice misfit between TiC and bcc iron is smaller than that between NbC and bcc iron. It is necessary, however, to treat more than 4000 atoms for the semi-coherent interface between TiC and bcc iron because of the small lattice misfit. The first-principles calculation of such a huge number of atoms is rather time-consuming even for materials with widely spread electron density. Moreover, the material used in this study includes the elements with localized electron density, namely, Ti, Fe, and C in comparison with a $3p$ electron system, such as Si. Thus, it is very difficult to treat this issue using the conventional first-principles calculation. In order to overcome this situation, we use the OpenMX package [24] in which the $O(N)$ Krylov-subspace method of first-principles calculation [25] is adopted. For the exchange-correlation energy in the generalized gradient approximation, the expression proposed by Perdew et al. is used [26]. The pseudopotential method with Troullier–Martins pseudopotential [27] is adopted. The $3p$, $3d$, and $4s$ states of Fe and Ti and $2s$ and $2p$ states of C are treated as valence electrons. The calculations are performed using double valence plus a single polarization function. In this study, the number of atoms in a truncated cluster is fixed at 64, 122, and 122 for Fe, Ti, and C, respectively, to ensure that the total energy differences between the $O(N)$ and conventional methods are less than 0.02 eV/atom. Since the most important quantity for the interface energy is the total energy, a convergence is checked by using the absolute error in total energy with respect to that of the conventional method.

For estimating the strain energy around the precipitate, structure optimization is performed using the empirical potential [28,29]. The first-principles calculation is known to give an accurate elastic constant with the experimental value [30]. The empirical potential is fitted to the experimental elastic constant. Thus, the compatibility between the first-principles calculation and the embedded-atom method calculation is fairly good. The matrix and precipitate regions are set in the simulation cell.

However, all atoms in the simulation cell are treated by the embedded-atom method potential for Fe [28,29]. In order to express the precipitate region, the atomic positions are fixed to have the same lattice parameters as TiC. In other words, we assume that the relaxation of atomic positions basically occurs in bcc iron. This assumption can be justified because, as mentioned earlier, the Young's modulus of TiC is approximately two times larger than that of bcc iron. The semi-coherent precipitate region is constructed by $2 \times (m \times m + (m-1) \times (m-1))$ atoms of Fe with an expansion of 0.7% along [001], and it is assumed to be $n \times n \times 3$ atoms of Ti or C. On the other hand, the coherent precipitate region is constructed by $2 \times (m \times m + (m-1) \times (m-1))$ atoms of Fe with expansions of 7.2% and 0.7% along [100] and [001], respectively, and it is assumed to be $m \times m \times 3$ atoms of Ti or C. The corresponding diameter of a precipitate is estimated by assuming precipitates to be spherical. In the matrix region, the atomic positions are optimized to reduce the force acting on each atom. The calculation cell is set to be 60^3 conventional unit cells to obtain strain energy accurately enough, because the strain energy due to a precipitate with the diameter of 2.8 nm is estimated in error by less than 0.01 J/m^2 between the 50^3 and 60^3 conventional unit cells.

4. Summary

The interface and the strain energies of the interface between the TiC and bcc iron are calculated using the $O(N)$ first-principles electronic structure calculation method and the structure optimization using the empirical potential, respectively. These calculations are performed for various precipitate sizes and coherencies. The interface and strain energies of the semi-coherent precipitate are compared with those of the coherent precipitate. As a result, the transition sizes from coherent to semi-coherent precipitate are estimated to be 1.5 and 2.3 nm for NbC and TiC, respectively. These precipitates undergo a transition from the coherent to semi-coherent state at sizes smaller than those with minimum misfit in order to release the large strain energies of the coherent precipitates. The transition size of TiC is larger than that of NbC, which is related to the smaller lattice misfit of TiC than that of NbC. The estimated transition size of TiC is close to the experimentally obtained size at which the change in the size dependence of the interaction force occurs. Therefore, the methodology presented herein provides a fairly accurate transition size of a precipitate from the coherent to semi-coherent state.

Acknowledgments: The numerical calculations were performed on the K computer provided by the RIKEN Advanced Institute for Computational Science through the HPCI System Research project (Project ID: hp120298, hp130016).

Author Contributions: Hideaki Sawada performed the calculations, analyzed the calculated data and wrote the paper. Hideaki Sawada, Shunsuke Taniguchi, Kazuto Kawakami and Taisuke Ozaki contributed the discussion for the analysis. Taisuke Ozaki developed the first-principles electronic structure simulation code OpenMX.

Abbreviations

The following abbreviations are used in this manuscript:

bcc	body-centered cubic
DOS	density of states
Fe–C–Ti	coherent interface where a C atom is connected with a Fe atom across the Fe–TiC interface
Fe–Ti–C	coherent interface where a Ti atom is connected with a Fe atom across the Fe–TiC interface
Fe–C–Nb	coherent interface where a C atom is connected with a Fe atom across the Fe–NbC interface
Fe–Nb–C	coherent interface where a Nb atom is connected with a Fe atom across the Fe–NbC interface

References

1. Maki, T. Possibilities of further increase in strength of steels. *Bull. Iron Steel Inst. Jpn.* **1998**, *3*, 781–786.

2. Yong, Q.; Sun, X.; Yang, G.; Zhang, Z. Solution and precipitation of secondary phase in steels: Phenomenon, Theory and Practice. In *Advanced Steels: The Recent Scenario in Steel Science and Technology*; Weng, Y., Dong, H., Gan Y., Eds.; Springer: New York, NY, USA, 2011; pp. 109–117.
3. Sudo, M.; Hashimoto, S.; Kobe, S. Niobium added ferrite-bainite high strength hot-rolled sheet steel with improved formability. *Tetsu-to-Hagané* **1982**, *68*, 1211–1220.
4. Funakawa, Y.; Shiozaki, T.; Tomita, K.; Yamamoto, T.; Maeda, E. Development of high strength hot-rolled sheet steel consisting of ferrite and nanometer-sized carbides. *ISIJ Int.* **2004**, *44*, 1945–1951.
5. Pressouyre, G.M.; Bernstein, I.M. Quantitative analysis of hydrogen trapping. *Metall. Trans.* **1978**, *9A*, 1571–1580.
6. Yamasaki, S.; Takahashi, T. Evaluation method of delayed fracture property of high strength steels. *Tetsu-to-Hagané* **1997**, *83*, 454–459.
7. Porter, D.A.; Easterling, K.E. *Phase Transformations in Metals and Alloys*; Taylor & Francis: London, UK, 1992; pp. 160–163.
8. Baker, R.G.; Nutting, J. The tempering of a Cr-Mo-V-W and a Mo-V steel. In *Precipitation Processes in Steels*; ISI Special Report No. 64; The Iron and Steel Institute: London, UK, 1959; pp. 1–22.
9. Wei, F.G.; Hara, T.; Tsuzaki, K. High resolution transmission electron microscopy study of crystallography and morphology of TiC precipitates in tempered steel. *Philos. Mag.* **2004**, *84*, 1735–1751.
10. Hartford, J. Interface energy and electron structure for Fe/VN. *Phys. Rev.* **2000**, *61*, 2221–2229.
11. Lee, J.H.; Shishidou, T.; Zhao, Y.J.; Freeman, A.J.; Olson, G.B. Strong interface adhesion in Fe/TiC. *Philos. Mag.* **2005**, *85*, 3683–3697.
12. Arya, A.; Carter, E.A. Structure, bonding, and adhesion at the TiC(100)/Fe(110) interface from first principles. *J. Chem. Phys.* **2003**, *118*, 8982–8996.
13. Chung, S.H.; Ha, H.P.; Jung, W.S.; Byun, J.Y. An ab initio study of the energies for interfaces between group V transition metal carbides and bcc iron. *ISIJ Int.* **2006**, *46*, 1523–1531.
14. Jung, W.S.; Chung, S.H.; Ha, H.P.; Byun, J.Y. An ab initio study of the energies for interfaces between group V transition metal nitrides and bcc iron. *Model. Simul. Mater. Sci. Eng.* **2006** *14*, 479–495.
15. Jung, W.S.; Chung, S.H.; Ha, H.P.; Byun, J.Y. An ab initio study of the energies of coherent interfaces formed between bcc iron and carbides or nitrides of transition metals. *Solid State Phenom.* **2007**, *124–126*, 1625–1628.
16. Jung, W.S.; Lee, S.C.; Chung, S.H. Energetics for interfaces between group IV transition metal carbides and bcc iron. *ISIJ Int.* **2008**, *48*, 1280–1284.
17. Jung, W.S.; Chung, S.H. Ab initio calculation of interfacial energies between transition metal carbides and fcc iron. *Model. Simul. Mater. Sci. Eng.* **2010**, *18*, 075008.
18. Johansson, S.A.E.; Christensen, M.; Wahnström, G. Interface energy of semicoherent metal-ceramic interfaces. *Phys. Rev. Lett.* **2005**, *95*, 226108.
19. Fors, D.H.R.; Wahnström, G. Theoretical study of interface structure and energetics in semicoherent Fe(001)/MX(001) systems. *Phys. Rev.* **2010**, *82*, 195410.
20. Fors, D.H.R.; Johansson, S.A.E.; Petisme, M.V.G.; Wahnström, G. Theoretical investigation of moderate misfit and interface energies in the Fe/VN system. *Comput. Mater. Sci.* **2010**, *50*, 550–559.
21. Sawada, H.; Taniguchi, S.; Kawakami, K.; Ozaki, T. First-principles study of interface structure and energy of Fe/NbC. *Model. Simul. Mater. Sci. Eng.* **2013**, *21*, 045012.
22. Taniguchi, S.; Shigesato, G. *Measurement of Coherent Strain Around TiC Precipitations in Ferrite*; Northwestern University—NIMS Materials Genome Workshop: Northwestern, IL, USA, 2012.
23. Kobayashi, Y.; Takahashi, J.; Kawakami, K. Experimental evaluation of the particle size dependence of the dislocation-particle interaction force in TiC-precipitation-strengthened steel. *Scr. Mater.* **2012**, *67*, 854–857.
24. The OpenMX Software Package. Available online: http://www.openmx-square.org/ (accessed on 17 July 2017).
25. Ozaki, T. O(N) krylov subspace method for large scale ab initio electronic structure calculations. *Phys. Rev.* **2006**, *74*, 245101.
26. Perdew, J.P.; Burke, K.; Ernzerhof, M. Generalized gradient approximation made simple. *Phys. Rev. Lett.* **1996**, *77*, 3865–3868.
27. Troullier, N.; Martins, J.L. Efficient pseudopotentials for plane-wave calculations. *Phys. Rev.* **1991**, *43*, 1993–2006.

28. Finnis, M.W.; Sinclair, J.E. A simple empirical n-body potential for transition metals. *Philos. Mag.* **1984**, *50*, 45–55.
29. Finnis, M.W.; Sinclair, J.E. Erratum. *Philos. Mag.* **1986**, *53*, 161.
30. Zhang, H.; Johansson, B.; Levente, V. Ab initio calculations of elastic properties of bcc Fe-Mg and Fe-Cr random alloys. *Phys. Rev.* **2009**, *79*, 224201.

metals

MDPI

Article

First-Principles Investigation of Structural, Electronic and Elastic Properties of HfX (X = Os, Ir and Pt) Compounds

Xianfeng Li [1,2], Cunjuan Xia [1], Mingliang Wang [1,*], Yi Wu [2] and Dong Chen [2,*]

[1] School of Materials Science and Engineering, Shanghai Jiao Tong University, Shanghai 200240, China; brucelee75cn@sjtu.edu.cn (X.L.); xiacunjuan@sjtu.edu.cn (C.X.)
[2] State Key Laboratory of Metal Matrix Composites, Shanghai Jiao Tong University, No. 800 Dongchuan Road, Shanghai 200240, China; eagle51@sjtu.edu.cn
[*] Correspondence: mingliang_wang@sjtu.edu.cn (M.W.); chend@sjtu.edu.cn (D.C.); Tel.: +86-21-5474-7597 (D.C.)

Received: 12 July 2017; Accepted: 10 August 2017; Published: 18 August 2017

Abstract: The structural, electronic and elastic properties of B2 structure Hafnium compounds were investigated by means of first-principles calculations based on the density functional theory within generalized gradient approximation (GGA) and local density approximation (LDA) methods. Both GGA and LDA methods can make acceptable optimized lattice parameters in comparison with experimental parameters. Therefore, both GGA and LDA methods are used to predict the electronic and elastic properties of B2 HfX (X = Os, Ir and Pt) compounds. Initially, the calculated formation enthalpies have confirmed the order of thermodynamic stability as HfPt > HfIr > HfOs. Secondly, the electronic structures are analyzed to explain the bonding characters and stabilities in these compounds. Furthermore, the calculated elastic properties and elastic anisotropic behaviors are ordered and analyzed in these compounds. The calculated bulk moduli are in the reduced order of HfOs > HfIr > HfPt, which has exhibited the linear relationship with electron densities. Finally, the anisotropy of acoustic velocities, Debye temperatures and thermal conductivities are obtained and discussed.

Keywords: Hf-based intermetallics; elastic properties; density functional theory; first-principles calculations

1. Introduction

Hafnium and Hafnium compounds can be used as the tools and parts in the nuclear power plants due to their high neutron cross sections [1], and medical implants and other medical industry applications owing to their excellent mechanical properties, high corrosive resistant ability and biocompatibility [2]. With the growing importance of Hafnium compounds, the studies on the theoretical and experimental aspects of these compounds have increased [3–9]. Levy et al. [3] performed a comprehensive study in the structural and thermodynamic properties of binary Hf compounds using ab initio calculations. Among them, the B2 structure is one of the most typical crystal structures for HfX compounds. For example, Novakovic et al. [4] studied the electronic structures, cohesive energies and formation enthalpies of B2 structure HfT_M (T_M = Co, Rh, Ru and Fe) compounds using ab initio full-potential linearized augmented plane waves calculations. Iyigör et al. [5] reported the structural, electronic, elastic and vibrational properties of HfX (X = Rh, Ru and Tc) using the plane-wave pseudopotential density functional theory via VASP codes. Xing et al. [6] investigated the structural phase stabilities of B2 phases and formation enthalpies of HfM (M = Ir, Os, Pd, Pt, Ru, Rh) using VASP codes. Wu et al. [7] studied the structural, elastic and electronic properties of HfRu compound theoretically. Although the structural features of B2 HfX (X = Os, Ir and Pt) compounds have been achieved discussed experimentally [10–12] and theoretically [6,13,14], the elastic and electronic properties of HfX (X = Os, Ir and Pt) compounds are rarely reported and assessed to our knowledge.

Therefore, this work has been organized as the following description. In Section 2, the computational methods of binary HfX (X = Os, Ir and Pt) compounds are presented. In Section 3, the results and discussions are exhibited and analyzed, including structural properties, electronic structures, elastic constants, elastic properties, elastic anisotropy, anisotropic sound velocities, Debye temperatures and thermal conductivities. In Section 4, the conclusions are made in detail.

2. Computational Information

The first-principles calculations based on the pseudopotential plane-wave within density functional theory (DFT) were performed using CASTEP (Cambridge Sequential Total Energy Package) codes [15,16]. The ultrasoft pseudopotential was adopted to simulate the ion-electron interaction [17]. Both generalized gradient approximation (GGA) with the Perdew-Burke-Ernzerhof (PBE) functional [18,19] and the local density approximation (LDA) functional with the form of Ceperley-Adler parameterized by Perdew and Zunger [20] were utilized to model the exchange-correlation. The basis sets have included atom states of $Hf5d^2 6s^2$, $Os5s^2 5p^6 5d^6 6s^2$, $Ir5d^7 6s^2$ and $Pt5d^9 6s^1$. With respect to cutoff energies and k-points, a series of convergence studies were performed. Afterwards, the cutoff energies were set at 400 eV. Besides, the special points sampling integration over the Brillouin zone was employed using Monkhorst-Pack method [21] with determined k-point separation of 0.02/Å in three lattice directions for each compound. Furthermore, the minimization scheme proposed by Brodyden-Fletcher-Goldfarb-Shanno (BFGS) was used during geometric optimization [22]. The tolerances of the geometrical optimization has to meet conditions, including the maximum ionic displacement $\leq 5.0 \times 10^{-4}$ Å, maximum ionic force ≤ 0.01 eV/Å, maximum stress ≤ 0.02 GPa, and the difference of total energy $\leq 5.0 \times 10^{-6}$ eV/atom. Followed by geometric optimization, the total energy and electronic structure were computed using self-consistent field tolerance of 5.0×10^{-7} eV/atom. Correspondingly, the lattice constants and atom coordinates were optimized via minimizing the total energy. At equilibrium structures, the corrected tetrahedron Blöchl method was utilized to derive the total energies [23].

3. Results and Discussion

3.1. Structural Properties

HfX (X = Os, Ir and Pt) compounds are in the (CsCl type) Pm3m space group, correspondingly. In an HfX unit cell, an Hf atom locates at 1a (0, 0, 0) and an X atom stays at 1b (0.5, 0.5, 0.5). In order to obtain structural properties of HfX compounds, the geometric optimizations are made firstly. The equilibrium lattice constants using both GGA and LDA methods are tabulated in Table 1, along with the available experimental [10–12] and theoretical [6,13,14] values for reference. Generally, the qualities for structural optimizations in HfX compounds using both GGA and LDA methods are comparable and acceptable. In detail, the optimized lattice constants for HfOs and HfIr are in good agreement with theoretical values [6,13,14], respectively. For HfOs and HfIr, the LDA computed values are more approaching to the experimental value. On the other aspect, the GGA method can give better optimized values than the LDA method in HfPt. Therefore, either method shows obvious superiority in the structural optimization of HfX compounds. Resultantly, both GGA and LDA methods are used in the following theoretical studies for HfX compounds.

Table 1. The optimized (a_{theo}) and experimental (a_{exp}) lattice constants, calculated deviations, bulk moduli (B_0) and its pressure derivatives (B_0'), and formation enthalpies (H_f) for HfX (X = Os, Ir and Pt).

Compounds	a_{theo} (Å)	a_{exp} (Å)	Calculated Deviation (%)	B_0 (GPa)	B_0'	H_f (eV/atom)
HfOs	3.291 [a]	3.239 [c]	1.600 [a]	234.3 [a]	4.43 [a]	−0.474 [a]
	3.236 [b]	-	−0.083 [b]	266.2 [b]	4.47 [b]	−0.451 [b]
	3.294 [d]	-	-	-	-	−0.495 [d]
	3.257 [e]	-	-	-	-	−0.707 [f]
	-	-	-	-	-	−0.793 [g]
	-	-	-	-	-	−0.484 ± −0.052 [h]
HfIr	3.311 [a]	3.21 [i]	3.143 [a]	220.9 [a]	4.66 [a]	0.807 [a]
	3.253 [b]	-	1.331 [b]	255.6 [b]	4.69 [b]	−0.769 [b]
	3.275 [k]	-	-	-	-	−0.977 [g]
	-	-	-	-	-	−1.016 ± −0.016 [j]
HfPt	3.359 [a]	3.3623 [k]	−0.096 [a]	191.6 [a]	4.64 [a]	−0.903 [a]
	3.298 [b]	-	−1.907 [b]	221.7 [b]	4.66 [b]	−0.900 [b]
	-	-	-	-	-	−1.063 [g]
	-	-	-	-	-	−1.175 ± −0.062 [j]

[a] Theoretical values from GGA method in current work; [b] Theoretical values from LDA method in current work; [c] Experimental values from reference [10]; [d] Theoretical values from reference [14]; [e] Theoretical values from reference [13]; [f] Theoretical values from reference [3]; [g] Theoretical values from reference [6]; [h] Experimental values from reference [24]; [i] Experimental values from reference [11]; [j] Experimental values from reference [25]; [k] Experimental values from reference [12].

Under increasing pressures from 0 to 25 GPa with per step of 5 GPa, the relative changes of (a) lattice constant and (b) unit cell volume for HfX compounds using GGA method are exhibited in Figure 1. In Figure 1a, the relative change of lattice constant (*a*) for HfPt is alternated larger than HfIr and HfOs under growing pressures. Similarly, the unit cell volume (V) changes larger in HfPt than HfIr and HfOs under pressures (Figure 1b). Therefore, the obtained pressure volume curve for each compound is formulated to a third-order Birch-Murnaghan equation of state (EOS) [26]:

$$P = \frac{3}{2}B_0[(\frac{V}{V_0})^{-\frac{7}{3}} - (\frac{V}{V_0})^{-\frac{5}{3}}]\{1 + \frac{3}{4}(B_0' - 4)[(\frac{V}{V_0})^{-\frac{2}{3}} - 1]\}. \tag{1}$$

For HfOs, HfIr and HfPt, the fitted bulk moduli (B_0) are 234.3 GPa, 220.9 GPa and 191.9 GPa, and their pressure derivatives (B_0') are 4.43, 4.66 and 4.64, accordingly. As a result, the bulk moduli are in the sequence of HfOs > HfIr > HfPt. Using LDA method, the bulk moduli are still in the order of HfOs > HfIr > HfPt, as exhibited in Table 1.

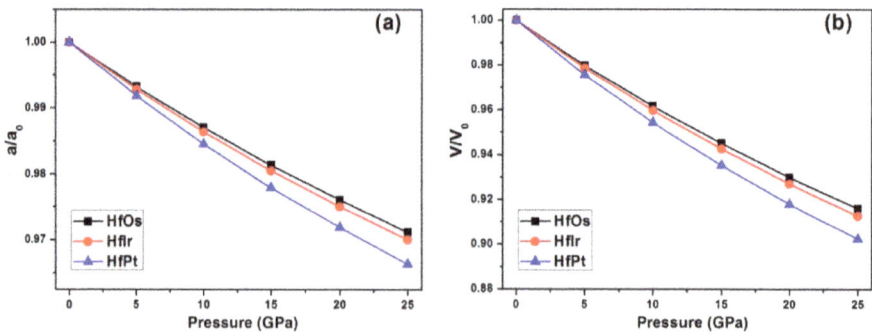

Figure 1. The relative changes of (**a**) lattice constant (*a*); (**b**) unit cell volume (*V*) with the elevating pressures computed by GGA method for HfX (X = Os, Ir and Pt).

The thermodynamic stability is closely associated with formation enthalpy in binary compounds. To evaluate the thermodynamic stability of a compound, the formation enthalpy (H_f) is expressed by the following equation [27,28]:

$$H_f^{HfX} = \frac{E_f^{HfX} - E_{Hf} - E_X}{2},$$ (2)

where E_f^{HfX} is the total energy of an HfX unit cell including an Hf atom and an X atom with equilibrium lattice parameters (X = Os, Ir and Pt); E_{Hf} and E_X are the total energy per atom of pure element solids at their ground states. Hf and Os are HCP metals, and Ir and Pt are FCC metals at ground state.

The negative formation enthalpy has denoted that the chemical process is exothermic, indicating the stability of resulted compounds. Moreover, the larger negative formation enthalpy has signified better stability of a compound [29]. In Table 1, the formation enthalpies of HfX compounds are tabulated. Generally, HfX compounds should be stable owing to their negative formation enthalpies derived from GGA and LDA methods. Furthermore, their thermodynamic stabilities are both in the order of HfPt > HfIr > HfOs. This conclusion is the same with Xing's work [6]. In detail, the H_f values calculated by GGA method are typically more negative than LDA method. For HfOs, the calculated H_f value by GGA method is in good consistency with experimental value [24] and theoretical value from Liu's work [14], but smaller than Xing's [6] and Levy's [3] reports. For HfIr and HfPt, our calculated H_f values are both smaller than theoretical values from Xing's work [6] and experimental values from Gachon's report [25]. Nevertheless, the crystal structures for HfIr and HfPt are not mentioned in this experimental work [25], which has degraded the reference value of this experimental work.

3.2. Density of States

The analyses on the total and partial density of states (DOS) are performed to further study the bonding characteristics and underlying mechanism of the structural stability of HfX compounds. Therefore, the total DOS (TDOS) and partial DOS (PDOS) computed by GGA method are presented and discussed herein. In Figure 2a, the Fermi level (E_F) is plotted at zero energy in all TDOS and PDOS spectra. Typically, there is not any energy gap identified near Fermi level for HfOs (Figure 2a), HfIr (Figure 2b) and HfPt (Figure 2c) in the TDOS, suggesting their essence of metallicity.

In the TDOS for HfOs (Figure 2a), the bonding interactions are dominated by the hybridization of Hfs and Oss states at -10 to -6 eV at the bonding states. Around the bonding states, the Osd states as the major role have intensively hybridized with the Hfd states as the minor role below the Fermi level. The two states have changed their roles in the hybridizations of above the Fermi level at the antibonding states. Around the Fermi level, a valley referring as a pseudogap, which is symbolized as the presence of covalent bonds [30,31], is considered to be the most observable feature in the TDOS. The pseudogaps have existed in the TDOS for HfOs (Figure 2a), HfIr (Figure 2b) and HfPt (Figure 2c), although their locations are different on energy scales. The pseudogap is located at the antibonding states for HfOs (Figure 2a), at the bonding states for HfIr (Figure 2b), and at the bonding states with the more negative energy states for HfPt (Figure 2c). Such differences should suggest HfOs has the least bonding stability.

For the sake of judging the structural stability of HfX compounds, the number of bonding electrons per atom is calculated based on the TDOS spectra. Since the charge interaction among bonding atoms is the critical factor to material's stability, the compound possessing higher number of bonding electrons should be more structurally stable [32–35]. For HfOs, HfIr and HfPt phases, the number of bonding electrons per atom are 6.013, 6.493 and 6.995, accordingly. Conclusively, the HfX phases have the stability order of HfPt > HfIr > HfOs. This conclusion is in good accordance with the thermodynamic analysis shown in Table 1.

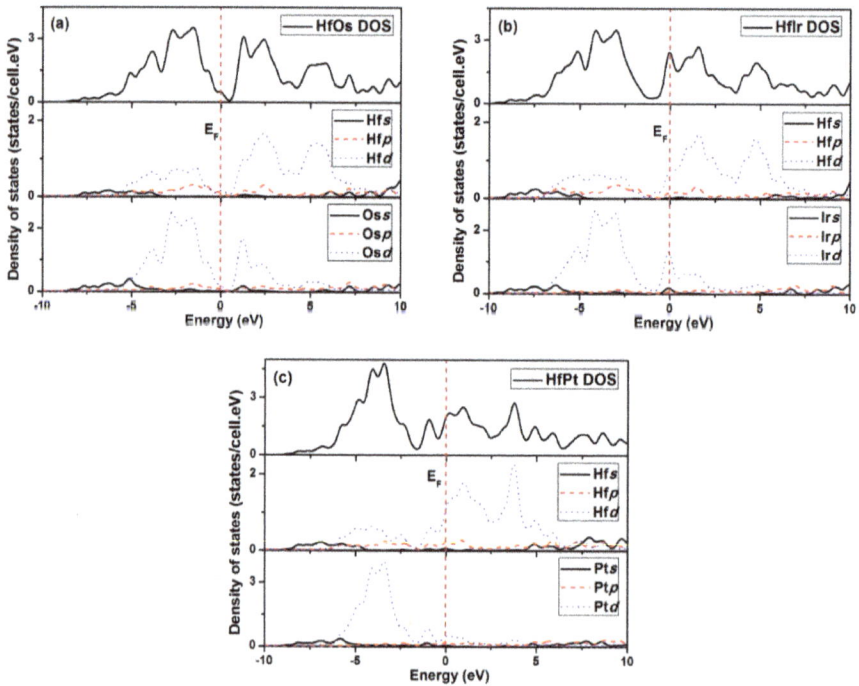

Figure 2. TDOS and PDOS spectra for (a) HfOs, (b) HfIr and (c) HfPt.

3.3. Elastic Properties

The elastic constants and elastic moduli are the critical information to study the mechanical properties of compounds. A full set of elastic constant for cubic crystal, including C_{11}, C_{12} and C_{44}, can be achieved using the stress-train method [36] (Table 2). Observably, the elastic constants obtained by LDA method are larger than GGA method in HfX compounds. In HfOs, the elastic constants derived from GGA method has agreed well with theoretical values from Arıkan's [13] and Liu's work [14]. Meanwhile, the elastic constants for pure Hf metal are calculated and presented with the published experimental values [37] for comparison, where the GGA computed elastic constants are in better agreement with experimental values [37].

In order to analyze the elastic properties of HfX compounds, the elastic constants generated using GGA method are used as the example. C_{11} is the symbol of compressive resistance along x axis. In each compound, the calculated C_{11} has the largest value, indicating the incompressible essence of the compound under the x direction uniaxial stress [38]. HfOs has the most incompressible ability owing to the largest C_{11} (402.1 GPa). Moreover, a larger C_{44} (121.0 GPa) can reflect a stronger resistance to monoclinic shear in (100) plane, suggesting HfOs also has the strongest ability to resist shear distortion in (100) plane.

Table 2. The elastic constants (C_{ij}) for HfX (X = Os, Ir and Pt).

Compounds	C_{11} (GPa)	C_{12} (GPa)	C_{13} (GPa)	C_{33} (GPa)	C_{44} (GPa)
HfOs	402.1 [a]	149.7 [a]	-	-	121.0 [a]
	436.5 [b]	179.3 [b]	-	-	130.5 [b]
	393.7 [c]	139.8 [c]	-	-	110.1 [c]
	366.1 [d]	152.6 [d]	-	-	105.6 [d]
HfIr	285.7 [a]	239.5 [a]	-	-	103.2 [a]
	255.7 [b]	202.9 [b]	-	-	90.7 [b]
HfPt	244.3 [a]	209.2 [a]	-	-	67.8 [a]
	217.7 [b]	179.3 [b]	-	-	61.7 [b]
Hf	214.0 [a]	90.6 [a]	92.2 [a]	235.8 [a]	54.6 [a]
	193.6 [b]	82.7 [b]	77.4 [b]	205.1 [b]	56.3 [b]
	181.0 [e]	77.0 [e]	66.0 [e]	197.0 [e]	55.7 [e]

[a] Theoretical values from GGA method in current work; [b] Theoretical values from LDA method in current work; [c] Theoretical values from reference [13]; [d] Theoretical values from reference [14]; [e] Experimental values from reference [37].

Furthermore, the mechanical stability is evaluated by Born's criteria [39] for cubic crystals:

$$C_{11} > 0; C_{44} > 0; C_{11} - C_{12} > 0; C_{11} + 2C_{12} > 0. \tag{3}$$

HfX (X = Os, Ir and Pt) compounds are all found mechanically stable at the ground state by the successful validation of Born's criteria.

In the engineering application, the elastic properties, i.e., bulk modulus (B), shear modulus (G), and Young's modulus (E), are demanded in practice. Generally, the elastic properties can be derived from Voigt-Reuss-Hill (VRH) methods using elastic constants [40]. For cubic crystals, the equations are expressed as following [27,41,42]:

$$B_V = B_R = \frac{1}{3}(C_{11} + 2C_{12}), \tag{4}$$

$$G_V = \frac{1}{5}(C_{11} - C_{12} + 3C_{44}), \tag{5}$$

$$G_R = \frac{5(C_{11} - C_{12})C_{44}}{4C_{44} + 3(C_{11} - C_{12})}, \tag{6}$$

$$B = \frac{B_V + B_G}{2}, \tag{7}$$

$$G = \frac{G_V + G_G}{2}, \tag{8}$$

As soon as the bulk and shear moduli are achieved, Young's modulus (E) and Poisson's ratio (ν) can be predicted:

$$E = \frac{9BG}{3B + G}, \tag{9}$$

$$\nu = \frac{3B - 2G}{2(3B + G)}, \tag{10}$$

The calculated elastic moduli, Poisson's ratio and B/G ratio using the VRH method at the ground state are tabulated for HfX compounds in Table 3, including the elastic moduli for pure Hf results for comparison. Overall, the elastic moduli obtained using GGA method are smaller than those provided by LDA method. Comparably, the calculated elastic moduli generated by the GGA method are in good agreement with theoretical values for HfOs. In detail, the elastic moduli are all larger than pure Hf in

HfOs. However, HfIr has the similar shear and Young's modulus over pure Hf, and HfPt owns the smaller values correspondingly. Therefore, HfOs should be a credible hardening phase in pure Hf.

Table 3. The bulk modulus (*B*), shear modulus (*G*), Young's modulus (*E*), Poisson's ratio (v), *B/G* ratio and hardness (*H*V) for HfX (X = Os, Ir and Pt) and pure Hf deduced from the VRH method.

Compounds	B (GPa)	G (GPa)	E (GPa)	v	Hv (GPa)	B/G
HfOs	233.8 [a]	123.1 [a]	314.1 [a]	0.276 [a]	13.4[a]	1.900 [a]
	265.1 [b]	129.7 [b]	334.6 [b]	0.290 [b]	12.8[b]	2.043 [b]
	224.4 [c]	116.5 [c]	298.0 [c]	0.279 [c]	12.7[c]	1.926 [c]
	223.8 [c]	106.1 [c]	274.8 [c]	0.295 [c]	10.7[c]	2.110 [c]
HfIr	220.5 [a]	55.5 [d]	153.5 [a]	0.384 [a]	3.3[a]	3.977 [a]
	254.9 [b]	57.2 [b]	159.6 [b]	0.396 [b]	3.0[b]	4.458 [b]
HfPt	192.1 [a]	38.7 [a]	108.8 [a]	0.406 [a]	2.0[a]	4.962 [a]
	220.9 [b]	39.6 [b]	112.2 [b]	0.415 [b]	1.8[b]	5.572 [b]
Hf	118.6 [a]	57.2 [a]	147.8 [a]	0.292 [a]	-	2.074 [a]
	134.8 [b]	59.9 [b]	156.5 [b]	0.306 [b]	-	2.251 [b]
	108.5 [e]	55.8 [e]	142.9 [e]	0.28 [e]	-	1.944 [e]

[a] Theoretical values from GGA method in current work; [b] Theoretical values from LDA method in current work; [c] Theoretical values from reference [13]; [d] Theoretical values from reference [14]; [e] Experimental values from reference [37].

Typically, the bulk modulus (*B*) suggests the resistant ability against volume change under pressure of materials. In addition, the bulk moduli calculated by both methods using VRH principles are in good agreement with those provided by the EOS equations (Table 1), confirming good self-consistency of this work. From Table 3, it is seen that HfOs possesses the largest resistance to volume change by applied pressure, while HfPt has the smallest. In order to illustrate the fundamental factor on bulk moduli of HfX compounds, the relationship between electron densities and bulk moduli are constructed in Figure 3a. Herein, the electron density (*n*) is the quotient of the bonding valence (Z_B) and the volume per atom (V_M) in metal [43]. For example, the electron density (*n*) is evaluated using the following equation in HfX compounds:

$$n(HfX) = Z_B(HfX)/V_M(HfX), \tag{11}$$

where $V_M(HfX)$ is the volume (cm^3/mol) of each HfX compound; $Z_B(HfX)$ is the bonding valence in (el/atom) rationalized from Vegard's law [44]:

$$Z_B(HfX) = (Z_B(Hf) + Z_B(X))/2, \tag{12}$$

where the bonding valence of pure element can be found in the reference [45].

In Figure 3a, the correlation between the electron density and bulk modulus can be constructed using the computed values either by LDA or GGA method. In detail, the linear relationships between the electron density and bulk modulus are clearly seen in HfX compounds.

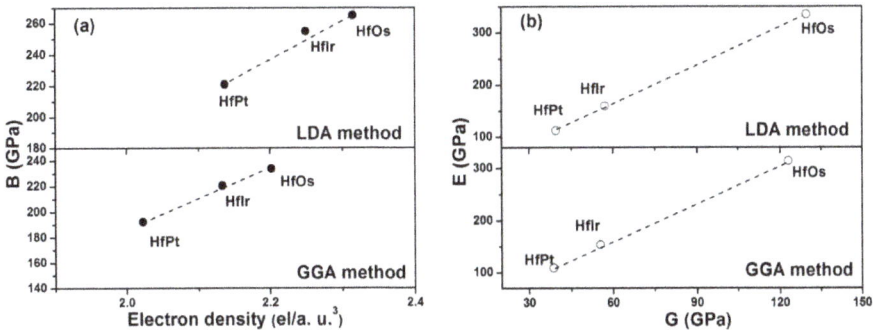

Figure 3. (a) Correlations between bulk modulus and electron density; (b) Correlations between shear modulus and Young's modulus for HfX (X = Os, Ir and Pt).

The shear modulus has reflected the resistance to reversible deformations under the shear stress [27]. A larger shear modulus for HfOs suggests its higher resistance to reversible deformations. Young's modulus is a symbol of the stiffness of a solid [46]. The material with a larger Young's modulus is stiffer. Therefore, HfOs is much stiffer than any other considered HfX compounds due to its higher Young's modulus. Overall, Young's modulus has linearly improved with the shear modulus in the order of HfOs > HfIr > HfPt for both LDA and GGA methods (Figure 3b).

The B/G value [47] and Poisson's ratio (v) [32,48] have determined the brittleness and ductility of the solid. A solid with $B/G < 1.75$ or $v < 0.26$ is usually brittle. Otherwise, it is ductile. In Table 3, it is found the HfPt is the most ductile compound with the largest B/G value and Poisson's ratio, and HfOs is the least ductile compound owing to the smallest B/G value and Poisson's ratio. Notably, the HfX compounds have similar ductile essence due to the small variations of B/G values and Poisson's ratios using both GGA and LDA methods.

The hardness (H_V) is associated with the plastic and elastic properties of an intermetallic compound. The hardness can be calculated by a semi-empirical equation [49]:

$$H_V = 0.92(G/B)^{1.137} G^{0.708},\tag{13}$$

where B and G are the bulk modulus and shear modulus, respectively.

The hardness values (H_V) of HfX (X = Os, Ir and Pt) compounds are tabulated in Table 3. Generally, the hardness is related with the values of G/B and G that the high values of G/B and G correspond to the high hardness. It is seen that the hardness value computed by GGA method is a bit larger than LDA method. Overall, the hardness values (H_V) have followed the order of HfPt < HfIr < HfOs in both GGA and LDA methods.

3.4. Elastic Anisotropy

The degree of elastic anisotropy is a critical property related to the engineering application. The universal anisotropic index (A^U) [50] is a universal measure to quantify the single crystalline elastic anisotropy in consideration of the contributions from both the bulk and the shear modulus, i.e., [51],

$$A^U = 5\frac{G_V}{G_R} + \frac{B_V}{B_R} - 6,\tag{14}$$

where B_V and G_V are the symbols of the Voigt bounds for bulk and shear modulus, respectively. B_R and G_R are the symbols of the Reuss bounds for bulk and shear modulus, respectively.

If $A^U = 0$ for a crystal, the crystal should be isotropic. A larger deviation of A^U from zero has indicated a severer degree of anisotropy. In cubic crystals, B_V/B_R is always equal to 1. Therefore, the universal anisotropic index is governed by G_V/G_R. Therefore, the calculated universal anisotropies by both LDA and GGA methods (Table 4) are reduced in the sequence of HfIr > HfPt > HfOs. In detail, HfIr and HfOs have the largest and smallest universal anisotropies, respectively.

Table 4. The calculated Voigt and Reuss bounds for bulk (shear) modulus, and universal anisotropic index (A^U) for HfX (X = Os, Ir and Pt).

Compounds	B_V	B_R	G_V	G_R	B_V/B_R	G_V/G_R	A^U
			GGA method				
HfOs	233.8	233.8	123.1	123.0	1	1.000	0.0022
HfIr	220.5	220.5	65.0	45.9	1	1.415	2.077
HfPt	192.1	192.1	44.7	32.7	1	1.367	1.830
			LDA method				
HfOs	265.1	265.1	129.7	129.7	1	1.000	0.00026
HfIr	254.9	254.9	71.1	43.2	1	1.646	3.229
HfPt	220.9	220.9	47.7	31.6	1	1.494	2.470

In order to describe the elastic anisotropic behavior more directly and effectively, the three-dimensional (3D) surface constructions of the directional dependence of reciprocal of Young's modulus have also been studied by following equations [52]:

$$\frac{1}{E} = S_{11} - 2(S_{11} - S_{12} - \frac{S_{44}}{2})(l_1^2 l_2^2 + l_2^2 l_3^2 + l_1^2 l_3^2), \tag{15}$$

where S_{ij} is the usual elastic compliance constant, which can be obtained from the inverse of the matrix of elastic constants; l_1, l_2 and l_3 are the direction cosines in the sphere coordination.

If the 3D directional dependence of Young's modulus exhibits a spherical shape, the crystal is ideal isotropic. Practically, the deviation extent from the spherical shape has reflected the degree of anisotropy. In Figure 4a, HfOs shows a quite spherical shape, signifying its near isotropic behavior. Furthermore, HfIr exhibits the largest deviation from the sphere shape with the strongest deviation along the <111> directions, confirming its intensive anisotropic behavior. Generally, the degree of the elastic anisotropy for HfX has followed the increasing order of HfOs < HfPt < HfIr. This conclusion is in good compliance with the result generated from the universal anisotropic index.

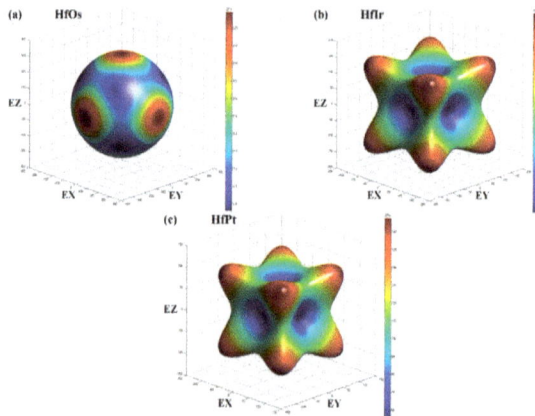

Figure 4. The 3D surface construction of Young's modulus in (**a**) HfOs, (**b**) HfIr and (**c**) HfPt by GGA method. (Magnitude of Young's modulus at different directions is presented by the contour along each graph with the unit of GPa.)

3.5. Anisotropy of Acoustic Velocities and Debye Temperature

In the crystalline material, the sound velocities are related to the crystal symmetry and the propagation direction. In cubic crystals, the pure transverse and longitudinal modes can be found in [111], [110] and [001] directions. With respect to other directions, the sound propagating modes have included the quasi-transverse or quasi-longitudinal waves. Therefore, the sound velocities can be derived from elastic constants using following expressions [53]:

$$[100]v_l = \sqrt{C_{11}/\rho};\ [010]v_{t1} = [001]v_{t2} = \sqrt{C_{44}/\rho}, \tag{16}$$

$$[110]v_l = \sqrt{(C_{11} + C_{12} + C_{44})/(2\rho)}, \tag{17}$$

$$[1\bar{1}0]v_{t1} = \sqrt{(C_{11} - C_{12})/\rho},\ [001]v_{t2} = \sqrt{C_{44}\rho}, \tag{18}$$

$$[111]_{vl} = \sqrt{(C_{11} + 2C_{12} + 4C_{44})/(3\rho)}, \tag{19}$$

$$[11\bar{2}]v_{t1} = [11\bar{2}]v_{t2} = \sqrt{(C_{11} - C_{12} + C_{44})/(3\rho)}, \tag{20}$$

where ρ is the density; v_l is the longitudinal sound velocity; v_{t1} and v_{t2} refer to the first and the second transverse mode of sound velocity, respectively.

Generally, C_{11} has determined the longitudinal sound velocity along [100] direction. C_{44} is related to the transverse modes along [010] and [001] directions. C_{11}, C_{12} and C_{44} can influence the longitudinal sound velocities along [110] and [111] directions in combination.

The longitudinal and the transverse sound velocities by both LDA and GGA methods along [100], [110] and [111] directions for HfX compounds are exhibited in Table 5. For the sound velocities obtained from LDA method, the longitudinal sound velocity v_l of each compound is reduced in the order of [111] > [110] > [100]. The anisotropic properties of sound velocities have also confirmed the elastic anisotropies in these cubic crystals. Meanwhile, the sound velocities derived from GGA method have shown the similar tendencies with those from LDA method.

Table 5. The calculated anisotropic sound velocities (m/s) for HfX (X = Os, Ir and Pt).

Crystal orientations	[111]		[110]			[100]		
	$[111]_{vl}$	$[11\bar{2}]_{vt1,2}$	$[110]_{vl}$	$[1\bar{1}0]_{vt1}$	$[001]_{vt2}$	$[100]_{vl}$	$[010]_{vt1}$	$[001]_{vt2}$
GGA method								
HfOs	4796.0	2691.9	4425.2	3833.6	2653.8	4838.3	2653.8	2653.8
HfIr	4487.1	1679.3	4024.3	1763.5	2313.0	3882.6	2313.0	2313.0
HfPt	4094.4	1427.9	3743.5	1531.8	1941.7	3647.2	1941.7	1941.7
LDA method								
IIfOs	4930.2	2674.7	4545.3	3773.4	2687.7	4916.0	2687.7	2687.7
HfIr	4684.4	1668.4	4191.3	1607.1	2401.7	3996.8	2401.7	2401.7
HfPt	4243.2	1408.4	3882.7	1424.5	1980.3	3758.8	1980.3	1980.3

The theoretically calculated structural properties (i.e., density) and elastic properties (i.e., Bulk modulus, shear modulus and Poisson's ratio) can be used to deduce Debye temperature (Θ), as shown in the following equation [26,54]:

$$\Theta = \frac{h}{k}\left[\frac{3n}{4\pi}\left(\frac{N_A\rho}{M}\right)\right]^{\frac{1}{3}} v_D, \tag{21}$$

where h is Planck's constant ($h = 6.626 \times 10^{-34}$ J/s); N_A is Avogadro's number ($N_A = 6.023 \times 10^{-23}$/mol); k is Boltzmann's constant ($k = 1.381 \times 10^{-23}$ J/K); M is the molecular weight; n is the number of atoms per formula unit.

v_D is the average sound velocity in polycrystalline materials, as exhibited using the equation:

$$v_D = [\frac{1}{3}(\frac{1}{v_L^3} + \frac{2}{v_T^3})]^{-\frac{1}{3}},$$

(22)

where v_T and v_L are the transverse and longitudinal sound velocities, including:

$$v_T = \sqrt{\frac{G}{\rho}},$$

(23)

$$v_L = \sqrt{\frac{B + \frac{4}{3}G}{\rho}},$$

(24)

Table 6 shows the sound velocity (m/s) and Debye temperature (K) for HfX (X = Os, Ir and Pt) calculated by GGA and LDA methods. Generally, these values calculated by LDA method have the larger values than those computed by GGA method. Specifically, the following reduced orders have shown as HfOs > HfIr > HfPt in Debye temperatures.

Table 6. Sound velocity (m/s) and Debye temperature (K) for HfX (X = Os, Ir and Pt).

Compounds	v_L (m/s)	v_T (m/s)	v_D (m/s)	Θ (K)
	GGA method			
HfOs	4812.7	2676.5	2978.3	339.5
HfIr	4166.6	1808.2	2040.8	231.2
HfPt	3859.0	1538.0	1741.1	194.4
	LDA method			
HfOs	4924.5	2679.9	2987.0	346.2
HfIr	4303.0	1788.1	2021.4	233.1
HfPt	3979.1	1514.3	1716.5	195.2

3.6. Thermal Conductivity

The thermal conductivities (k) is a useful physical parameter for practical applications. The thermal conductivity is reduced with elevating temperature to a limiting value known as the minimum thermal conductivity (k_{min}), which can be evaluated according to Cahill's model [55]:

$$k_{min} = \frac{k}{2.48}n^{\frac{2}{3}}(v_l + v_{t1} + v_{t2}),$$

(25)

where k is Boltzmann constant; n is the number of density of atoms per volume. v_l and v_t are the longitudinal and transverse sound velocities, respectively (Table 7).

Table 7 exhibits the calculated k_{min} values using Cahill's model. Generally, the lattice thermal conductivity and the electronic thermal conductivity are the main compositions to the total thermal conductivity. At lower temperature, the effect from electron-phonon scattering is considered limited. Therefore, the thermal conductivities of HfX (X = Os, Ir and Pt) are ascribed to the lattice thermal conductivities at the ground state. The derived thermal conductivities are found small thereby. Based on Callaway-Debye theory [56], the lattice thermal conductivity is proportional to Debye temperature. It means a higher Debye temperature should correspond to a larger lattice thermal conductivity. Therefore, HfOs has the larger thermal conductivity than HfIr and HfPt in order. It is seen that the values offered by the GGA method are always smaller, but own the similar tendency with the LDA method. These HfX compounds show the relatively lower thermal conductivities, indicating that they are poor thermal conductors at the ground state.

Cahill's model is suitable to discuss the anisotropic behavior of compounds on the thermal conductivity, since it has involved with treating the total thermal conductivity in association with each acoustic branch. For instance, it is seen that the calculated thermal conductivities have exhibited anisotropic behaviors owing to the differences of v_l, v_{t1} and vt2 along [100], [110] and [111] directions. In detail, k_{min}[111] is always smaller than k_{min}[100] and k_{min}[110], suggesting that the dependence of thermal conductivities along [111] direction is less prominent than those along [100] and [110] directions. Although there is lacking experimental values for comparison, our theoretical results should prove the guidance for future studies.

Table 7. Calculated minimum thermal conductivities k_{min} (W/m/K) for HfX (X = Os, Ir and Pt).

Compounds	n (10^{303})	[100] k_{min}	[110] k_{min}	[111] k_{min}	k_{min}
		GGA method			
HfOs	0.0561	0.828	0.890	0.391	0.829
HfIr	0.0551	0.686	0.653	0.362	0.627
HfPt	0.0528	0.590	0.565	0.321	0.543
		LDA method			
HfOs	0.0590	0.868	0.929	0.416	0.868
HfIr	0.0581	0.735	0.685	0.391	0.658
HfPt	0.0558	0.627	0.592	0.345	0.569

4. Conclusions

The structural, electronic and elastic properties of B2 HfX (X = Os, Ir and Pt) compounds have been studied using first-principles calculations. Initially, the structural optimizations are comparable and acceptable for HfX compounds using both GGA and LDA methods. The calculated formation enthalpies by GGA and LDA methods have confirmed that the thermodynamic stability is in the order of HfPt > HfIr > HfOs. Secondly, the calculated electronic structures are derived, and similar features are identified in DOS spectra for HfX. The results show the sequence of structural stability should be HfPt > HfIr > HfOs, which is in good agreement with thermodynamic analyses. Mechanically, the elastic moduli obtained by GGA method are typically smaller than LDA method. In detail, the calculated bulk moduli using VRH method are in good agreement with those provided by EOS equation. Besides, the calculated bulk moduli are in the order of HfOs > HfIr > HfPt, where the bulk moduli can be correlated with electron densities in compounds. Additionally, Young's modulus has augmented linearly with the shear modulus. HfOs has the Young's modulus of 314.1 GPa and hardness of 13.4 GPa by GGA method, which should be a credible hardening phase in pure Hf. Then, the ductile essence is in the sequence of HfPt > HfIr > HfOs according to the analyses on Poisson's ratio and B/G ratio. Based the universal anisotropic indexes and 3D surface constructions, the elastic anisotropy has followed the increasing order of HfIr > HfPt > HfOs. Finally, the anisotropy of acoustic velocities, Debye temperatures and thermal conductivities are obtained and discussed. Our results are hoped to inspire future experimental and theoretical investigations on these Hf-based compounds.

Acknowledgments: This work is sponsored by the Research Fund (Project No. 15X100040018) at Shanghai Jiao Tong University, China.

Author Contributions: Xianfeng Li has conducted on the literature search, data collection, data analysis, data interpretation and writing the manuscript. Cunjuan Xia has conducted on the data collection, figure preparation and commenting on the manuscript. Mingliang Wang has conducted on the literature search, study design, data interpretation and revising the manuscript. Yi Wu has conducted on the data analysis, figure preparation and commenting on the manuscript. Dong Chen has conducted on the study design, data analysis, data interpretation and revising the manuscript.

Conflicts of Interest: The authors declare no conflict of interest.

References

1. Wallenius, J.; Westlén, D. Hafnium clad fuels for fast spectrum BWRs. *Ann. Nucl. Energy* **2008**, *35*, 60–67. [CrossRef]
2. Herranz-Diez, C.; Mas-Moruno, C.; Neubauer, S.; Kessler, H.; Gil, F.J.; Pegueroles, M. tuning mesenchymal stem cell response onto titanium-niobium-hafnium alloy by recombinant fibronectin fragments. *ACS Appl. Mater. Interfaces* **2016**, *8*, 2517–2525. [CrossRef] [PubMed]
3. Levy, O.; Hart, G.L.; Curtarolo, S. Hafnium binary alloys from experiments and first principles. *Acta Mater.* **2010**, *58*, 2887–2897. [CrossRef]
4. Novaković, N.; Ivanović, N.; Koteski, V.; Radisavljević, I.; Belošević-Čavor, J.; Cekić, B. Structural stability of some CsCl structure HfT$_M$ (T$_M$ = Co, Rh, Ru, Fe) compounds. *Intermetallics* **2006**, *14*, 1403–1410. [CrossRef]
5. İyigör, A.; Özduran, M.; Ünsal, M.; Örnek, O.; Arıkan, N. Ab-initio study of the structural, electronic, elastic and vibrational properties of HfX (X = Rh, Ru and Tc). *Philos. Mag. Lett.* **2017**, *97*, 110–117. [CrossRef]
6. Xing, W.; Chen, X.Q.; Li, D.; Li, Y.; Fu, C.L.; Meschel, S.V. First-principles studies of structural stabilities and enthalpies of formation of refractory intermetallics: TM and TM$_3$ (T = Ti, Zr, Hf; M = Ru, Rh, Pd, Os, Ir, Pt). *Intermetallics* **2012**, *28*, 16–24. [CrossRef]
7. Wu, J.; Liu, S.; Zhan, Y.; Yu, M. Ternary addition and site substitution effect on B2 RuHf-based intermetallics: A first-principles study. *Mater. Des.* **2016**, *108*, 230–239. [CrossRef]
8. Gueorguiev, G.K.; Pacheco, J.M. Silicon and metal nanotemplates: Size and species dependence of structural and electronic properties. *J. Chem. Phys.* **2003**, *119*, 10313. [CrossRef]
9. Gueorguiev, G.K.; Pacheco, J.M.; Stafström, S.; Hultman, L. Silicon-metal clusters: Nano-templates for cluster assembled materials. *Thin Solid Films* **2006**, *515*, 1192. [CrossRef]
10. Dwight, A.E. CsCl-type equiatomic phases in binary alloys of transition elements. *Trans. AIME* **1959**, *215*, 283–286.
11. Korniyenko, K.Y.; Kriklya, L.S.; Khoruzhaya, V.G. The Hf-Ir-Ru (Hafnium-Iridium-Ruthenium). *J. Ph. Equilib. Diffus.* **2014**, *35*, 369–376. [CrossRef]
12. Stalick, J.K.; Waterstrat, R.M. The hafnium-platinum phase diagram. *J. Ph. Equilib. Diffus.* **2014**, *35*, 15–23. [CrossRef]
13. Arıkan, N.; Örnek, O.; Charifi, Z.; Baaziz, H.; Uğur, Ş.; Uğur, G. A first-principle study of Os-based compounds: Electronic structure and vibrational properties. *J. Phys. Chem. Solids* **2016**, *96–97*, 121–127. [CrossRef]
14. Liu, Q.J.; Zhang, N.C.; Liu, F.S.; Liu, Z.T. Structural, mechanical and electronic properties of OsTM and TMOs$_2$ (TM = Ti, Zr and Hf): First-principles calculations. *J. Alloys Compd.* **2014**, *589*, 278–282. [CrossRef]
15. Segall, M.D.; Lindan, P.J.; Probert, M.J.; Pickard, C.J.; Hasnip, P.J.; Clark, S.J. First-principles simulation: ideas, illustrations and the CASTEP code. *J. Phys. Condens. Matter* **2002**, *14*, 2717–2744. [CrossRef]
16. Clark, S.J.; Segall, M.D.; Pickard, C.J.; Hasnip, P.J.; Probert, M.J.; Refson, K. First principles methods using CASTEP. *Z. Kristallogr.* **2005**, *220*, 567–570. [CrossRef]
17. Vanderbilt, D. Soft self-consistent pseudopotentials in a generalized eigenvalue formalism. *Phys. Rev. B* **1990**, *41*, 7892. [CrossRef]
18. Perdew, J.P.; Wang, Y. Accurate and simple analytic representation of the electron-gas correlation energy. *Phys. Rev. B* **1992**, *45*, 13244. [CrossRef]
19. Perdew, J.P.; Burke, K.; Ernzerhof, M. Generalized gradient approximation made simple. *Phys. Rev. Lett.* **1996**, *77*, 3865. [CrossRef] [PubMed]
20. Perdew, J.P.; Zunger, A. Self-interaction correction to density-functional approximations for many-electron systems. *Phys. Rev. B* **1981**, *23*, 5048–5079. [CrossRef]
21. Shanno, D.F. Conditioning of quasi-Newton methods for function minimization. *Math. Comput.* **1970**, *24*, 647–656. [CrossRef]
22. Fischer, T.H.; Almlof, J. General Methods for Geometry and Wave Function Optimization. *J. Chem. Phys.* **1992**, *96*, 9768–9774. [CrossRef]
23. Blöchl, P.E.; Jepsen, O.; Andersen, O.K. Improved tetrahedron method for Brillouin-zone integrations. *Phys Rev. B* **1994**, *49*, 16223. [CrossRef]
24. Mahdouk, K.; Gachon, J.C. Calorimetric study of the Hf-Os and Os-Ti system. *J. Alloys Compd.* **1998**, *278*, 185–189. [CrossRef]

25. Gachon, J.C.; Selhaoui, N.; Aba, B.; Hertz, J. Comparison between measured and predicted enthalpies of formation. *J. Ph. Equilib.* **1992**, *13*, 506–511. [CrossRef]

26. Zhong, S.Y.; Chen, Z.; Wang, M.; Chen, D. Structural, elastic and thermodynamic properties of Mo_3Si and Mo_3Ge. *Eur. Phys. J. B* **2016**, *89*, 6. [CrossRef]

27. Chen, D.; Chen, Z.; Wu, Y.; Wang, M.; Ma, N.; Wang, H. First-principles investigation of mechanical, electronic and optical properties of Al_3Sc intermetallic compound under pressure. *Comput. Mater. Sci.* **2014**, *91*, 165–172. [CrossRef]

28. Chen, D.; Xia, C.; Chen, Z.; Wu, Y.; Wang, M.; Ma, N. Thermodynamic, elastic and electronic properties of $AlSc_2Si2$. *Mater. Lett.* **2015**, *138*, 148–150. [CrossRef]

29. Shang, X.; Shen, J.; Tian, F. A first-principles study of the tetragonal and hexagonal R_2Al (R = Cr, Zr, Nb, Hf, Ta) phases. *Mater. Res. Express* **2016**, *3*, 106503. [CrossRef]

30. Krajci, M.; Hafner, J. Covalent bonding and bandgap formation in intermetallic compounds: A case study for Al_3V. *J. Phys. Condens Matter* **2002**, *14*, 1865–1879. [CrossRef]

31. Xu, J.H.; Freeman, A.J. Bandfilling and structural stability of trialuminides: YAl_3, $ZrAl_3$, and $NbAl_3$. *J. Mater. Res.* **1991**, *6*, 1188–1199. [CrossRef]

32. Hu, W.C.; Liu, Y.; Li, D.J.; Zeng, X.Q.; Xu, C.S. First-principles study of structural and electronic properties of C14-type Laves phase Al_2Zr and Al_2Hf. *Comput. Mater. Sci.* **2014**, *83*, 27–34. [CrossRef]

33. Hou, H.; Wen, Z.; Zhao, Y.; Fu, L.; Wang, N.; Han, P. First-principles investigations on structural, elastic, thermodynamic and electronic properties of Ni_3X (X = Al, Ga and Ge) under pressure. *Intermetallics* **2014**, *44*, 110–115. [CrossRef]

34. Hu, W.C.; Liu, Y.; Li, D.J.; Zeng, X.Q.; Xu, C.S. Mechanical and thermodynamic properties of Al_3Sc and Al_3Li precipitates in Al-Li-Sc alloys from first-principles calculations. *Physica B* **2013**, *427*, 85–90. [CrossRef]

35. Kong, Y.; Duan, Y.; Ma, L.; Li, R. Phase stability, elastic anisotropy and electronic structure of cubic MAl_2 (M = Mg, Ca, Sr and Ba) Laves phases from first-principles calculations. *Mater. Res. Express* **2016**, *3*, 106505. [CrossRef]

36. Ding, W.J.; Yi, J.X.; Chen, P.; Li, D.L.; Peng, L.M.; Tang, B.Y. Elastic properties and electronic structures of typical Al-Ce structures from first-principles calculations. *Solid State Sci.* **2012**, *14*, 555–561. [CrossRef]

37. Chen, Q.; Sundman, B. Calculation of debye temperature for crystalline structures—A case study on Ti, Zr, and Hf. *Acta Mater.* **2001**, *49*, 947–961. [CrossRef]

38. Gao, X.; Jiang, Y.; Zhou, R.; Feng, J. Stability and elastic properties of Y-C binary compounds investigated by first principles calculations. *J. Alloys Compd.* **2014**, *587*, 819–826. [CrossRef]

39. Born, M.; Huang, K. *Dynamical Theory of Crystal Lattices*; Oxford University Press: Oxford, UK, 1954.

40. Hill, R. The Elastic Behaviour of a Crystalline Aggregate. *Phys. Soc. Lond. Sect. A* **1952**, *65*, 349–354. [CrossRef]

41. Voigt, W. *Lehrbuchde Kristallphysik*; B.G. Teubner: Leipzig/Berlin, Germany, 1928.

42. Reuss, A. Calculation of the flow limits of mixed crystals on the basis of the plasticity of monocrystals. *Z. Angew. Math. Mech.* **1929**, *9*, 49–58. [CrossRef]

43. Huang, S.; Zhang, C.H.; Li, R.Z.; Shen, J.; Chen, N.X. Site preference and alloying effect on elastic properties of ternary B2 RuAl-based alloys. *Intermetallics* **2014**, *51*, 24–29. [CrossRef]

44. Jacob, K.T.; Raj, S.; Rannesh, L. Vegard's law: A fundamental relation or an approximation? *Int. J. Mater. Res.* **2007**, *98*, 776. [CrossRef]

45. Li, C.; Wu, P. Correlation of Bulk Modulus and the Constituent Element Properties of Binary Intermetallic Compounds. *Chem. Mater.* **2001**, *13*, 4642. [CrossRef]

46. McNaught, A.D.; Wilkinson, A. Compendium of Chemical Terminology. In *The Gold Book*, 2nd ed.; Blackwell Scientific Publications: Oxford, UK, 1997; Volume 2, pp. 12–14, ISBN 0-86542-684-8.

47. Pugh, S.F. XCII. Relations between the elastic moduli and the plastic properties of polycrystalline pure metals. *Philos. Mag* **1954**, *45*, 823–843. [CrossRef]

48. Greaves, G.N.; Greer, A.L.; Lakes, R.S.; Rouxel, T. Poisson's ratio and modern materials. *Nat. Mater.* **2011**, *10*, 823–837. [CrossRef] [PubMed]

49. Tian, Y.; Xu, B.; Zhao, Z. Microscopic theory of hardness and design of novel superhard crystals. *Int. J. Refract. Met. Hard Mater.* **2012**, *33*, 93–106. [CrossRef]

50. Lv, Z.Q.; Zhang, Z.F.; Zhang, Q.; Wang, Z.H.; Sun, S.H.; Fu, W.T. Structural, electronic and elastic properties of the Laves phases WFe$_2$, MoFe$_2$, WCr$_2$ and MoCr$_2$ from first-principles. *Solid State Sci.* **2016**, *56*, 16–22. [CrossRef]

51. Duan, Y.H.; Huang, B.; Sun, Y.; Peng, M.J.; Zhou, S.G. Stability, elastic properties and electronic structures of the stable Zr-Al intermetallic compounds: A first-principles investigation. *J. Alloys Compd.* **2014**, *590*, 50–60. [CrossRef]

52. Chen, S.; Sun, Y.; Duan, Y.H.; Huang, B.; Peng, M.J. Phase stability, structural and elastic properties of C15-type Laves transition-metal compounds MCo$_2$ from first-principles calculations. *J. Alloys Compd.* **2015**, *630*, 202–208. [CrossRef]

53. Duan, Y.H.; Sun, Y.; Peng, M.J.; Zhou, S.G. Anisotropic elastic properties of the Ca-Pb compounds. *J. Alloys Compd.* **2014**, *595*, 14–21. [CrossRef]

54. Vajeeston, P.; Ravindran, P.; Fjellvag, H. Prediction of structural, lattice dynamical, and mechanical properties of CaB$_2$. *RSC Adv.* **2012**, *2*, 11687–11694. [CrossRef]

55. Cahill, D.G.; Watson, S.K.; Pohl, R.O. Lower limit to the thermal conductivity of disordered crystals. *Phys Rev. B* **1992**, *46*, 6131. [CrossRef]

56. Duan, Y.; Sun, Y. Thermodynamics properties and thermal conductivity of Mg$_2$Pb at high pressure. *Sci. China Phys. Mech. Astron.* **2013**, *56*, 1854–1860. [CrossRef]

metals MDPI

Article

Molecular Dynamics Simulation of Crack Propagation in Nanoscale Polycrystal Nickel Based on Different Strain Rates

Yanqiu Zhang and Shuyong Jiang *

College of Mechanical and Electrical Engineering, Harbin Engineering University, Harbin 150001, China; zhangyq@hrbeu.edu.cn
* Correspondence: jiangshuyong@hrbeu.edu.cn; Tel.: +86-0451-8251-9710

Received: 11 September 2017; Accepted: 3 October 2017; Published: 16 October 2017

Abstract: Based on the strain rates of 2×10^8 s^{-1} and 2×10^{10} s^{-1}, molecular dynamics simulation was conducted so as to study mechanisms of crack propagation in nanoscale polycrystal nickel. The strain rate has an important effect on the mechanism of crack propagation in nanoscale polycrystal nickel. In the case of a higher strain rate, local non-3D-crystalline atoms are induced and Lomer-Cottrell locks are formed, which plays a critical role in crack initiation and propagation. Orientation difference between adjacent grains leads to the slipping of dislocations along the different directions, which results in the initiation of a void near the triple junction of grain boundaries and further contributes to accelerating the crack propagation.

Keywords: metals; deformation; crack growth; polycrystal; molecular dynamics

1. Introduction

Fracture of engineering materials frequently leads to failure of structures and components and thus this issue has attracted more and more attention in the engineering fields. Continuum mechanics and theory have played a significant role in investigating the mechanisms of fracture of engineering materials. In particular, fracture frequently takes place along with plastic deformation in metal materials. In general, the occurrence of fracture is closely related to the initiation and propagation of cracks or the initiation and growth of voids. In the most common circumstances, the mechanism of crack formation in the ductile metals involves a process of nucleation, growth and coalescence of voids. Therefore, it is of great significance to reveal the mechanism of initiation and propagation of cracks on the basis of several length scales, such as the macroscale, mesoscale, microscale and nanoscale (or atomic scale). So far, based on the aforementioned four scales, many researchers have devoted themselves to revealing the mechanisms of initiation and propagation of cracks by numerical simulation methods which principally include continuum mechanics finite element method (CMFEM), crystal plasticity finite element method (CPFEM), discrete dislocation dynamics (DDD) method and molecular dynamics (MD) method. These simulation methods are suitable for various scales, where each scale can be separately used for analyzing the mechanisms of crack propagation. Furthermore, multiscale modeling and simulation have been applied to investigate the mechanisms of crack propagation. One approach is to deliver information from a lower scale to a higher scale by means of the parameters involved. Another approach is to directly couple the different scales on the basis of a single computation, such as coupling molecular dynamics with continuum plasticity [1], coupling discrete dislocation dynamics with crystal plasticity [2], and coupling molecular dynamics with discrete dislocation dynamics [3].

It is well known that the simulation methods based on different scales play different roles in analyzing the mechanisms of crack propagation. In general, CMFEM is established on the basis of linear elastic or elastic plastic fracture mechanics. Therefore, the merit of CMFEM is that it is able to resolve the fracture problems at the macroscopic scale, where it can be used to obtain the stress field, the strain field, the size and shape of plastic zone around the crack tip and the stress intensity factor [4,5]. Compared with CMFEM, the main advantage of CPFEM is that it is able to tackle anisotropic micromechanical problems in crystalline materials, where their mechanical effects are dependent on orientation due to the underlying crystalline structure [6,7]. Therefore, CPFEM becomes a powerful candidate for understanding the mechanisms of crack propagation when the crack propagates in a single crystal or in a polycrystal with crystallographic texture [8,9]. Kartal et al. [10] studied the effect of crystallographic orientation on stress state and plasticity at the crack tip in a hexagonal close-packed (HCP) single crystal Ti through CPFEM and they found that the size and shape of the plastic zone at the crack tip is strongly dependent on crystallographic orientation, but the stress state at the crack tip shows little dependence on crystallographic orientation. Sabnis et al. [11] studied the influence of secondary orientation on plastic deformation near the notch tip in a notched Ni-based single crystal by means of CPFEM, where the slip patterns and the size and shape of plastic zone are strongly dependent on the secondary orientation of the notch, which has an important effect on the initiation and propagation of crack. Li et al. [12] predicted the initiation of a fatigue crack in a polycrystalline aluminum alloy by means of CPFEM, where the grain size, grain shape and grain orientation are considered. Lin et al. [13] studied the propagation of a transgranular crack in a polycrystalline nickel-based superalloy by means of CPFEM and they demonstrated that plastic deformation at the crack tip is related to the grain orientation.

Discrete dislocations play a dual role in plastic deformation at a crack tip. On one hand, plastic deformation, which results from the motion of dislocations, enhances the resistance to crack propagation. On the other hand, local stress concentrations induced by organized dislocation structures contribute to crack propagation. Therefore, the advantage of the DDD method is that it is able to simulate the plastic deformation near the crack tip or around the void according to the evolution of discrete dislocations and hence it is capable of capturing the formation of dislocation structures at a microscopic scale [14–16]. In particular, the DDD method has been frequently used to capture the details of discrete dislocation evolution near the crack tip and around the void. So far, the DDD method has been applied from a single crystal material to a polycrystalline material. Segurado and Llorca [17] revealed the micromechanisms of plastic deformation and void growth in an isolated face-centered cubic (FCC) single crystal by means of the DDD method. Liang et al. [18] investigated the interaction between the blunt crack and the void in a single crystal by virtue of the DDD method, and they focused on the crack-tip deformation, the void growth. Huang et al. [19] applied the DDD method to simulating plastic deformation near a transgrannular crack tip under cyclic loading conditions by taking into account both dislocation climb and dislocation-grain boundary in a polycrystalline nickel-based superalloy.

As an atomistic simulation technique, MD simulation plays an important role in describing the atomic behavior during plastic deformation and crack propagation of metal materials under mechanical loading. So far, MD simulation has been extensively applied to studying the basic mechanism of crack propagation in single crystal and bicrystal metal materials. Sung and Chen [20] studied crack propagation in single crystal nickel via MD simulation, where they found that Shockley partial dislocations are initiated at the crack tip, and they then propagate along the close-packed (111) plane until the crystal is fractured. Zhang and Ghosh [21] investigated the deformation mechanisms at the crack tip of single crystal nickel with an embedded crack by means of MD simulation and they proposed that some lattice orientations lead to the emission of dislocation loops, whereas other lattice orientations contribute to the formation of twins. Zhou et al. [22] investigated the intergranular crack propagation behaviors of copper bicrystals with different symmetrical tilt grain boundaries via MD simulation. Shimokawa and Tsuboi [23] studied the atomic-scale mechanism of plasticity at intergranular crack tip of aluminum bicrystal with tilt grain boundary via MD simulation. Zhang et al. [24] investigated the mechanism for intercrystalline crack propagation at twist boundary of nanoscale bicrystal nickel by means of MD.

To our knowledge, few papers in the literature have reported the mechanisms of crack propagation in the polycrystal metal materials via MD simulation. In the present study, on the basis of MD simulation, we comparatively investigate the mechanisms of crack propagation in nanoscale polycrystal nickel in the case of the different strain rates in order to reveal the effect of strain rate on the initiation and propagation of a crack.

2. Modeling and Methods

2.1. Simulation Model

The MD models for nanoscale polycrystal nickel is established based on the non-periodic boundary conditions, as shown in Figure 1. In the simulation, [111], [$\bar{1}\bar{1}$2] and [1$\bar{1}$0] directions were set as *x*, *y* and *z* directions of the Cartesian reference frame, respectively. In addition, twist boundaries were applied in all the grain boundaries. The lattice orientations for the grains are illustrated in Figure 1a. All the models were determined as the dimension of 20 × 20 × 5 unit cells, and a notch with the size of 5 × 1 × 5 unit cells was preset in one grain. The simulation is performed according to the following two steps. Firstly, energy minimization was conducted on the initial configurations by fixing the y-direction boundary. Secondly, a velocity-controlled tensile load was applied on the two ends labeled fixed atoms in Figure 1a, where the strain rates were chosen as 2×10^8 s^{-1} and 2×10^{10} s^{-1}, respectively. All the simulations were performed at 0.01 K in order to eliminate the thermal effects. Figure 1b,c illustrate the configurations before relaxation and after relaxation, respectively.

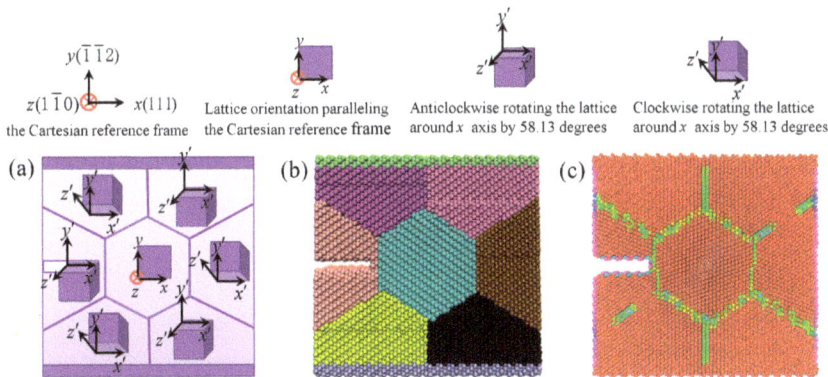

Figure 1. Models for nanoscale polycrystal nickel: (**a**) diagrams of grain orientations; (**b**) configurations before relaxation; (**c**) configurations after relaxation.

2.2. Potential Function

In the current investigation, in order to describe the interactions among nickel atoms, the embedded-atom-method (EAM) potential proposed by Foiles et al. [25] was applied, as expressed in Equation (1).

$$E_{total} = \sum_i F_i(\rho_{h,i}) + \frac{1}{2}\sum_i \sum_{j\neq i} \varphi_{ij}(R_{ij}) \tag{1}$$

where E_{total} is the total energy in the system, $\rho_{h,i}$ the host electron density at atom *i* due to the remaining atoms of the system, $F_i(\rho)$ the energy for embedding atom *i* into the background electron density ρ, and $\varphi_{ij}(R_{ij})$ the core-core pair repulsion between atoms *i* and *j* separated by the distance R_{ij}. In addition,

F_i is related to the element of atom i alone and φ_{ij} is related to the elements of atoms i and j. As stated above, the electron density is approximated by the superposition of atomic densities, namely

$$\rho_{h,i} = \sum_{j \neq i} \rho_j(R_{ij}) \tag{2}$$

where $\rho_j(R_{ij})$ is the electron density supplied by atom j.

2.3. Simulation Methods

In the current investigation, all the simulations were implemented via LAMMPS software [26]. Since only the global mechanical behaviors are considered, Virial stress for an atomic system was used. Lagrangian virial stress has been confirmed to be the most accurate one in terms of evaluating the stress value in dynamical simulations [26].

The centrosymmetry parameter (CSP) was used for analyzing the Shockley partial dislocations as well as stacking faults (SFs), which are induced by tensile deformation. As for the FCC structure, the CSP is defined as follows [27].

$$P = \sum_{i=1,6} \left| \vec{R}_i + \vec{R}_{i+6} \right|^2 \tag{3}$$

where R_i and R_{i+6} are the vectors which correspond to six pairs of opposite nearest neighbors in the FCC lattice. As for atoms in a perfect nickel lattice, the CSP, P, is defined as zero. However, if the nickel lattice is subjected to distortion, the value of P will not be zero. Instead, the specific value of the parameter P will correspond to a particular defect. The dislocation atoms can be observed when all the surface and perfect atoms within the simulation domain are removed.

In order to visualize the defects in the nanoscale nickel crystals, atoms were colored via common neighbor analysis (CNA) [28]. This method is able to detect the environment which an atom is located in, HCP environment or FCC environment. AtomEye [29], a visualization software, was applied for identifying the deformation mechanisms during simulations. In the current investigation, HCP atoms were displayed by light blue color, FCC atoms were expressed by dark blue color and non-3D-crystalline atoms were exhibited by red color. In addition, VMD (Version 1.9.3, Theoretical and Computational Biophysics Group, Urbana, IL, USA, 2016), another visualization software, was also applied to identify the changes in grain boundaries [30].

3. Results and Discussion

Figures 2 and 3 show the process of crack propagation, along with plastic deformation, in nanoscale polycrystal nickel in the case of 2×10^8 s^{-1} and 2×10^{10} s^{-1}, respectively. It can be found from Figures 2 and 3 that in the process of crack propagation, a void begins to occur in the interior of the middle grain. With the progression of plastic deformation, the void gradually grows and finally merges with the initial crack, which contributes to the rapid crack propagation. As a consequence, at the strain of 0.48, the middle grain is subjected to fracture. In the case of high strain rate, in particular, the void tends to be nucleated at trigeminal grain boundaries. In addition, as for different strain rates, the direction in which the void grows exhibits a certain distinction. The phenomenon indicates that the strain rate has an important effect on mechanism of crack propagation in nanoscale polycrystal nickel. It can be generally accepted that the strain rate plays a critical role in the mechanical properties of crystal since it has a predominant influence on the crack propagation. In order to further understand the mechanical properties of nanoscale polycrystal nickel, tensile stress-strain curves are obtained, as shown in Figure 4. It can be seen that purely elastic deformation occurs at early stage of deformation, where the stress increases linearly with increasing strain. When the strain reaches a critical value, the stress drops sharply with increasing strain, which is aroused by the rapid crack propagation. Furthermore, it is evident that the yield stress of nanoscale polycrystal nickel at the strain rate of

2×10^{10} s^{-1} is much higher compared with the one at the strain rate of 2×10^8 s^{-1}, which means that the crack propagation is sensitive to the strain rate.

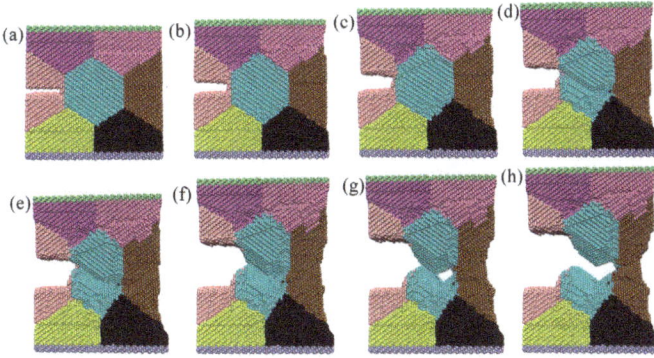

Figure 2. Process of crack propagation in the polycrystal nickel at the strain rate of 2×10^8 s^{-1}: (**a**) $\varepsilon = 0$; (**b**) $\varepsilon = 0.06$; (**c**) $\varepsilon = 0.12$; (**d**) $\varepsilon = 0.18$; (**e**) $\varepsilon = 0.24$; (**f**) $\varepsilon = 0.36$; (**g**) $\varepsilon = 0.42$; (**h**) $\varepsilon = 0.48$.

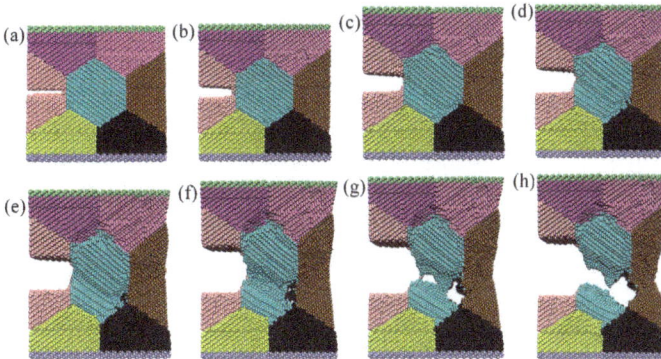

Figure 3. Process of crack propagation in the polycrystal nickel at the strain rate of 2×10^{10} s^{-1}: (**a**) $\varepsilon = 0$; (**b**) $\varepsilon = 0.06$; (**c**) $\varepsilon = 0.12$; (**d**) $\varepsilon = 0.18$; (**e**) $\varepsilon = 0.24$; (**f**) $\varepsilon = 0.36$; (**g**) $\varepsilon = 0.42$; (**h**) $\varepsilon = 0.48$.

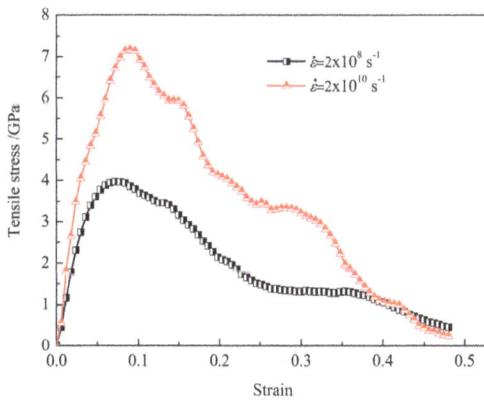

Figure 4. Tensile stress-strain curves for the polycrystal nickel at different strain rates.

For the purpose of further clarifying the process of crack propagation in nanoscale polycrystal nickel, the evolution of the atomic configuration along with crack propagation is captured on the basis of CNA in the case of 2×10^8 s^{-1} and 2×10^{10} s^{-1}, as shown in Figures 5 and 6. It can be found that the dislocations which are emitted from the grain boundaries and the crack tip are capable of moving along the slip plane in the various directions because there is the orientation difference among the grains. In the case of the two strain rates, dislocation slip mainly appears in the middle grain, as well as in the grain which is located to the right of the middle grain. Furthermore, the directions in which the dislocations slip in the two grains exhibit the great discrepancy, which is responsible for the initiation of the void. Compared to 2×10^8 s^{-1}, HCP atoms are distributed in a more complicated way in the nanoscale polycrystal nickel subjected to tensile deformation at the strain rate of 2×10^{10} s^{-1}, where more slip planes are activated.

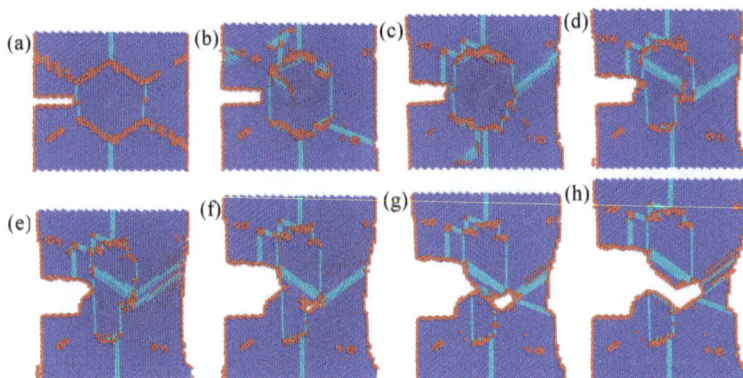

Figure 5. Evolution of atomic configuration for crack propagation in polycrystal nickel at the strain rate of 2×10^8 s^{-1}: (**a**) $\varepsilon = 0$; (**b**) $\varepsilon = 0.06$; (**c**) $\varepsilon = 0.12$; (**d**) $\varepsilon = 0.18$; (**e**) $\varepsilon = 0.24$; (**f**) $\varepsilon = 0.36$; (**g**) $\varepsilon = 0.42$; (**h**) $\varepsilon = 0.48$. (All the configurations are characterized by common neighbor analysis (CNA), where hexagonal close-packed (HCP) atoms were displayed by light blue color, face-centered cubic (FCC) atoms were expressed by dark blue color and non-3D-crystalline atoms were exhibited by red color).

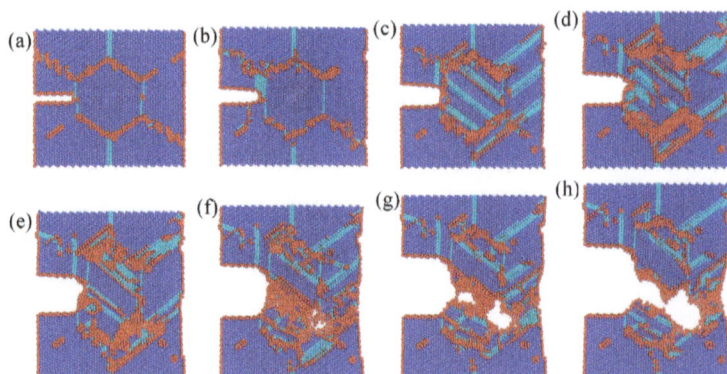

Figure 6. Evolution of atomic configuration for crack propagation in polycrystal nickel at the strain rate of 2×10^{10} s^{-1}: (**a**) $\varepsilon = 0$; (**b**) $\varepsilon = 0.06$; (**c**) $\varepsilon = 0.12$; (**d**) $\varepsilon = 0.18$; (**e**) $\varepsilon = 0.24$; (**f**) $\varepsilon = 0.36$; (**g**) $\varepsilon = 0.42$; (**h**) $\varepsilon = 0.48$. (All the configurations are characterized by CNA, where HCP atoms were displayed by light blue color, FCC atoms were expressed by dark blue color and non-3D-crystalline atoms were exhibited by red color).

For the sake of further clarifying the mechanism of crack propagation in the nanoscale polycrystal nickel, the evolution of the atomic configuration, along with the crack propagation, is captured on the basis of CSP, as shown in Figures 7 and 8. It can be observed from Figure 7 that, with the progression of plastic deformation, Shockley partial dislocations occur along with stacking faults. In particular, at the tensile strain of 0.18, the stacking faults in the adjacent grains are arranged against the grain boundary in the different directions. This phenomenon further indicates that the dislocations in the different grains slip along the different directions in order to accommodate the plastic deformation due to the existence of the grain boundary. It can be seen in Figure 8 that at the early stage of plastic deformation, two Shockley partial dislocations with a Burgers vector $b_1 = 1/6\,[11\bar{2}]\,(111)$ are initiated from the grain boundary and the crack tip, respectively. Subsequently, increasing Shockley partial dislocations are emitted from the grain boundaries and move along the close-packed $\{111\}$ planes in the grains. It can be generally accepted that the polycrystal possesses more defects due to the existence of the grain boundaries. As a consequence, more dislocations are emitted from the grain boundaries and then they advance along the various directions due to the orientation difference among the grains.

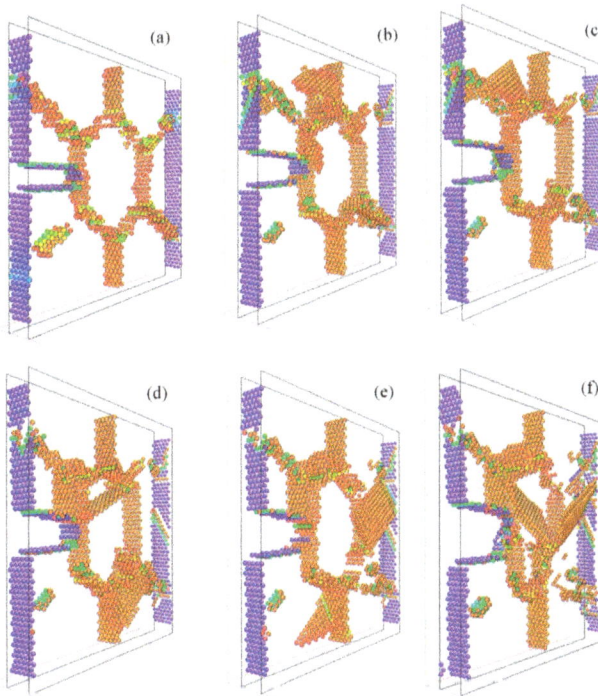

Figure 7. Evolution of atomic configuration in nanoscale polycrystal nickel subjected to early deformation at the strain rate of $2 \times 10^8\ \text{s}^{-1}$: (**a**) $\varepsilon = 0$; (**b**) $\varepsilon = 0.06$; (**c**) $\varepsilon = 0.084$; (**d**) $\varepsilon = 0.096$; (**e**) $\varepsilon = 0.12$; (**f**) $\varepsilon = 0.18$. (All the configurations are characterized by centrosymmetry parameter (CSP), where stacking faults are companied by Shockley partial dislocations).

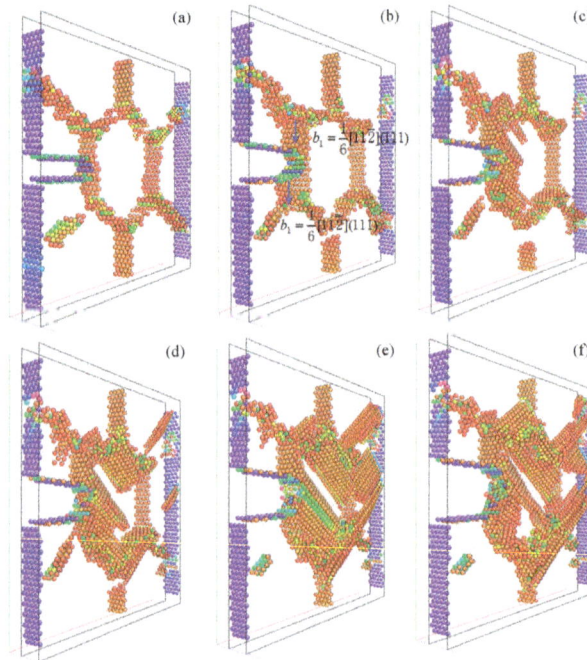

Figure 8. Evolution of atomic configuration in nanoscale polycrystal nickel subjected to early deformation at the strain rate of 2×10^{10} s^{-1}: (**a**) $\varepsilon = 0$; (**b**) $\varepsilon = 0.06$; (**c**) $\varepsilon = 0.084$; (**d**) $\varepsilon = 0.096$; (**e**) $\varepsilon = 0.12$; (**f**) $\varepsilon = 0.18$. (All the configurations are characterized by CSP, where stacking faults are companied by Shockley partial dislocations).

Figures 9 and 10 give a more detailed description of the evolution of the atomic configuration along with crack propagation in the nanoscale polycrystal nickel on the basis of CSP in the case of 2×10^8 s^{-1} and 2×10^{10} s^{-1}, respectively. It can be found that when the nanoscale polycrystal nickel is subjected to plastic deformation in the case of 2×10^{10} s^{-1}, local non-3D-crystalline atoms are induced and Lomer-Cottrell locks are formed. However, in the case of 2×10^8 s^{-1}, neither non-3D-crystalline atoms nor Lomer-Cottrell locks are induced, but stacking faults are captured. The phenomena indicate that the mechanisms of crack propagation are different from each other in the case of the two different strain rates. It can be generally accepted that Lomer-Cottrell locks are induced by the interaction between two Shockley partial dislocations which move along the two adjacent {111} slip planes. The Lomer-Cottrell locks are capable of impeding the movement of dislocations, which results in pile-ups of dislocations. As a consequence, the Lomer-Cottrell locks play a critical role in the initiation and propagation of cracks. With the progression of plastic strain, more and more non-3D-crystalline atoms are induced near the Lomer-Cottrell lock. Eventually, the non-3D-crystalline atoms are separated from one another so as to relax the deformation energy, which consequently leads to the initiation and propagation of a crack. Compared to the strain rate of 2×10^{10} s^{-1}, when the nanoscale polycrystal nickel experiences plastic deformation in the case of 2×10^8 s^{-1}, the less slip systems are activated and consequently Lomer-Cottrell locks are difficult to be formed and simultaneously the non-3D-crystalline atoms are not easy to be induced as well. It can be concluded that in the case of the two strain rates, the initiation of void is more attributed to the fact that the orientation difference between the two adjacent grains leads to the slip of the dislocations along the different directions. Therefore, the occurrence of the void is unavoidable so as to accommodate the compatibility of deformation.

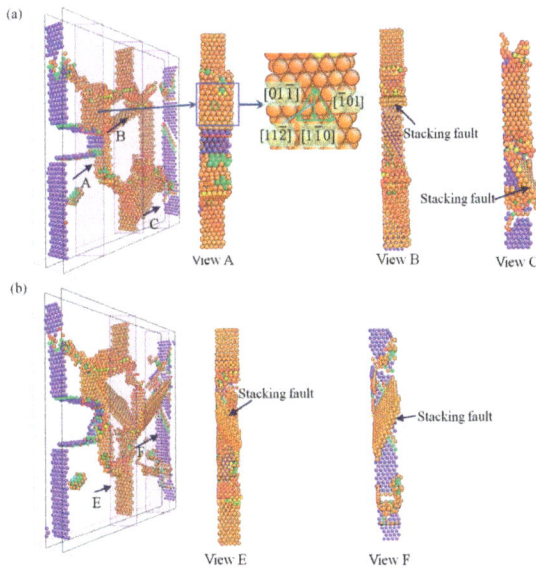

Figure 9. Detailed evolution of atomic configuration in nanoscale polycrystal nickel at the strain rate of 2×10^8 s^{-1}: (**a**) $\varepsilon = 0.096$; (**b**) $\varepsilon = 0.18$. (All the configurations are characterized by CSP, where stacking faults are companied by Shockley partial dislocations).

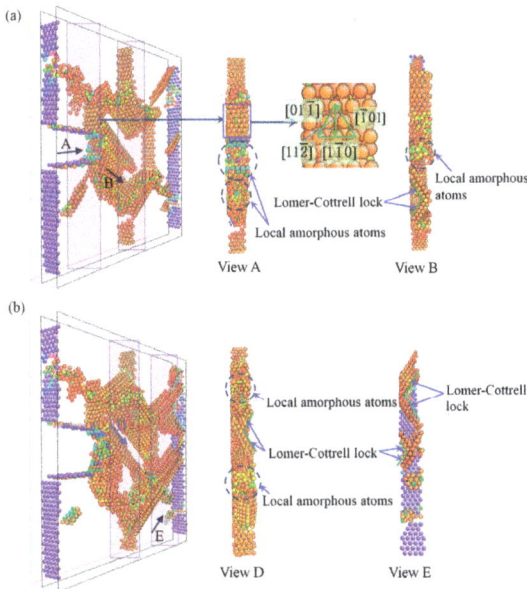

Figure 10. Detailed evolution of atomic configuration in nanoscale polycrystal nickel at the strain rate of 2×10^{10} s^{-1}: (**a**) $\varepsilon = 0.096$; (**b**) $\varepsilon = 0.18$. (All the configurations are characterized by CSP, where stacking faults are companied by Shockley partial dislocations).

4. Conclusions

Based on the strain rates of 2×10^8 s^{-1} and 2×10^{10} s^{-1}, the mechanisms of crack propagation in nanoscale polycrystal nickel are investigated by means of molecular dynamics simulation. The following conclusions are drawn.

(1) The strain rate has an important effect on mechanisms of crack propagation in nanoscale polycrystal nickel. The yield stress is sensitive to the strain rate, and the yield stress of nanoscale polycrystal nickel at the strain rate of 2×10^{10} s^{-1} is much higher than the counterpart at the strain rate of 2×10^8 s^{-1}.

(2) When the nanoscale polycrystal nickel is subjected to plastic deformation in the case of 2×10^{10} s^{-1}, local non-3D-crystalline atoms are induced and Lomer-Cottrell locks are formed. However, in the case of 2×10^8 s^{-1}, neither non-3D-crystalline atoms nor Lomer-Cottrell locks are induced, but stacking faults are captured.

(3) Lomer-Cottrell locks are able to impede the movement of dislocations and thus lead to the pile-up of dislocations. In addition, the non-3D-crystalline atoms are easy to be induced near the Lomer-Cottrell locks. Consequently, Lomer-Cottrell locks contribute to accelerating the propagation of crack.

(4) The orientation difference between the adjacent grains leads to slipping of dislocations along the different directions, which results in the initiation of a void in the vicinity of the triple junction of grain boundaries. The coalescence of the void with the crack contributes to accelerating the propagation of cracks.

Acknowledgments: The work was financially supported by National Natural Science Foundation of China (No. 51475101 and 51305091).

Author Contributions: Yanqiu Zhang, performed MD simulation and wrote the manuscript; Shuyong Jiang, supervised the research.

Conflicts of Interest: The authors declare no conflict of interest.

References

1. Yamakov, V.I.; Warner, D.H.; Zamora, R.J.; Saether, E.; Curtin, W.A.; Glaessgen, E.H. Investigation of crack tip dislocation emission in aluminum using multiscale molecular dynamics simulation and continuum modeling. *J. Mech. Phys. Solids* **2014**, *65*, 35–53. [CrossRef]
2. Cui, Y.; Liu, Z.; Zhuang, Z. Quantitative investigations on dislocation based discrete-continuous model of crystal plasticity at submicron scale. *Int. J. Plasticity* **2015**, *65*, 54–72. [CrossRef]
3. Shilkrot, L.; Miller, R.; Curtin, W.A. Multiscale plasticity modeling: Coupled atomistics and discrete dislocation mechanics. *J. Mech. Phys. Solids* **2004**, *52*, 755–787. [CrossRef]
4. Han, Q.; Wang, Y.; Yin, Y.; Wang, D. Determination of stress intensity factor for mode I fatigue crack based on finite element analysis. *Eng. Fract. Mech.* **2015**, *138*, 118–126. [CrossRef]
5. Benz, C.; Sander, M. Reconsiderations of fatigue crack growth at negative stress ratios: Finite element analyses. *Eng. Fract. Mech.* **2015**, *145*, 98–114. [CrossRef]
6. Ghosh, S.; Anahid, M. Homogenized constitutive and fatigue nucleation models from crystal plasticity FE simulations of Ti alloys, Part 1: Macroscopic anisotropic yield function. *Int. J. Plasticity* **2013**, *47*, 182–201. [CrossRef]
7. Ghosh, S.; Anahid, M. Homogenized constitutive and fatigue nucleation models from crystal plasticity FE simulations of Ti alloys, Part 2: Macroscopic probabilistic crack nucleation model. *Int. J. Plasticity* **2013**, *48*, 111–124.
8. Proudhon, H.; Li, J.; Wang, F.; Roos, A.; Chiaruttini, V.; Forest, S. 3D simulation of short fatigue crack propagation by finite element crystal plasticity and remeshing. *Int. J. Fatigue* **2016**, *82*, 238–246. [CrossRef]
9. Li, J.; Proudhon, H.; Roos, A.; Chiaruttini, V.; Forest, S. Crystal plasticity finite element simulation of crack growth in single crystals. *Comp. Mater. Sci.* **2014**, *94*, 191–197. [CrossRef]

10. Kartal, M.E.; Cuddihy, M.A.; Dunne, F.P.E. Effects of crystallographic orientation and grain morphology on crack tip stress state and plasticity. *Int. J. Fatigue* **2014**, *61*, 46–58. [CrossRef]
11. Sabnis, P.A.; Mazière, M.; Forest, S.; Arakere, N.K.; Ebrahimi, F. Effect of secondary orientation on notch-tip plasticity in superalloy single crystals. *Int. J. Plasticity* **2012**, *28*, 102–123. [CrossRef]
12. Li, L.; Shen, L.; Proust, G. Fatigue crack initiation life prediction for aluminium alloy 7075 using crystal plasticity finite element simulations. *Mech. Mater.* **2015**, *81*, 84–93. [CrossRef]
13. Lin, B.; Zhao, L.G.; Tong, J. A crystal plasticity study of cyclic constitutive behaviour, crack-tip deformation and crack-growth path for a polycrystalline nickel-based superalloy. *Eng. Fract. Mech.* **2011**, *78*, 2174–2192. [CrossRef]
14. Huang, M.; Zhao, L.; Tong, J.; Li, Z. Discrete dislocation dynamics modelling of mechanical deformation of nickel-based single crystal superalloys. *Int. J. Plasticity* **2012**, *28*, 141–158. [CrossRef]
15. Quek, S.S.; Wu, Z.; Zhang, Y.W.; Srolovitz, D.J. Polycrystal deformation in a discrete dislocation dynamics framework. *Acta. Mater.* **2014**, *75*, 92–105. [CrossRef]
16. Fan, H.; Aubry, S.; Arsenlis, A.; El-Awady, J.A. The role of twinning deformation on the hardening response of polycrystalline magnesium from discrete dislocation dynamics simulation. *Acta Mater.* **2015**, *92*, 126–139. [CrossRef]
17. Segurado, J.; Llorca, J. Dsicrete dislocation dynamics analysis of the effect of lattice orientation on void growth in single crystals. *Int. J. Plasticity* **2010**, *26*, 806–819. [CrossRef]
18. Liang, S.; Huang, M.; Li, Z. Discrete dislocation modeling on interaction between type-I blunt crack and cylindrical void in single crystals. *Int. J. Solids Struct.* **2015**, *56–57*, 209–219. [CrossRef]
19. Huang, M.; Tong, J.; Li, Z. A study of fatigue crack tip characteristics using discrete dislocation dynamics. *Int. J. Plasticity* **2014**, *54*, 229–246. [CrossRef]
20. Sung, P.H.; Chen, T.C. Studies of crack growth and propagation of single-crystal nickel by molecular dynamics. *Comp. Mater. Sci.* **2015**, *102*, 151–158. [CrossRef]
21. Zhang, J.; Ghosh, S. Molecular dynamics based study and characterization of deformation mechanisms near a crack in a crystalline material. *J. Mech. Phys. Solids* **2013**, *61*, 1670–1690. [CrossRef]
22. Zhou, Y.; Yang, Z.; Lu, Z. Dynamic crack propagation in copper bicrystals grain boundary by atomistic simulation. *Mater. Sci. Eng. A* **2014**, *599*, 116–124. [CrossRef]
23. Shimokawaa, T.; Tsuboi, M. Atomic-scale intergranular crack-tip plasticity in tilt grain boundaries acting as an effective dislocation source. *Acta Mater.* **2015**, *87*, 233–247. [CrossRef]
24. Zhang, Y.; Jiang, S.; Zhu, X.; Zhao, Y. A molecular dynamics study of intercrystalline crack propagation in nano-nickel bicrystal films with (0 1 0) twist boundary. *Eng. Fract. Mech.* **2016**, *168*, 147–159. [CrossRef]
25. Foiles, S.M.; Baskes, M.I.; Daw, M.S. Embedded-atom-method functions for the FCC metals Cu, Ag, Au, Ni, Pd, Pt, and their alloys. *Phys. Rev. B* **1986**, *33*, 7983–7991. [CrossRef]
26. Plimpton, S.; Parallel, F. Fast parallel algorithms for short-range molecular dynamics. *J. Comput. Phys.* **1995**, *117*, 1–19. [CrossRef]
27. Kelchner, C.L.; Plimpton, S.J.; Hamilton, J.C. Dislocation nucleation and defect structure during surface indentation. *Phys. Rev. B* **1998**. [CrossRef]
28. Tsuzuki, H.; Branicio, P.S.; Rino, J.P. Structural characterization of deformed crystals by analysis of common atomic neighborhood. *Comput. Phys. Commun.* **2007**, *177*, 518–523. [CrossRef]
29. Li, J. AtomEye: An efficient atomistic configuration viewer. *Model. Simul. Mater. Sci. Eng.* **2003**, *11*, 173–177. [CrossRef]
30. Humphrey, W.; Dalke, A.; Schulten, K. VMD-Visual Molecular Dynamics. *J. Mol. Graph.* **1996**, *14*, 33–38. [CrossRef]

metals

MDPI

Article

Ab Initio-Based Modelling of the Yield Strength in High-Manganese Steels

Simon Sevsek * and Wolfgang Bleck

Steel Institute, RWTH Aachen University, Intzestraße 1, 52072 Aachen, Germany; bleck@iehk.rwth-aachen.de
* Correspondence: simon.sevsek@iehk.rwth-aachen.de; Tel.: +49-241-80-90138

Received: 15 November 2017; Accepted: 3 January 2018; Published: 5 January 2018

Abstract: An ab initio-based model for the strength increase by short-range ordering of C-Mn-Al clusters has been developed. The model is based on ab initio calculations of ordering energies. The impact of clusters on the yield strength of high-manganese austenitic steels (HMnS) is highly dependent on the configurational structure of the cells that carbon atoms will position themselves as interstitial atoms. The impact of the alloying elements C, Mn, and Al on the potential and actual increase in yield strength is analyzed. A model for the calculation of yield strengths of HMnS is derived that includes the impact of short-range ordering, grain size refinement, and solid solution strengthening. The model is in good agreement with experimental data and performs better than other models that do not include strengthening by short-range ordering.

Keywords: short-range ordering; yield strength; high-manganese steel

1. Introduction

High Mn austenitic steels (HMnS) offer great combinations of hardness and ductility with a wide range of possible applications in various industries, such as the automotive industry [1–3]. Depending on the stacking fault energy, which is a result of alloying content, temperature, deformation grade, and other factors, HMnS exhibit a variety of different active deformation modes [4–7]. Those deformation modes include transformation induced plasticity (TRIP), twinning-induced plasticity (TWIP) or pronounced planar glide in combination with the formation of dislocation cells that is sometimes referred to as microband-induced plasticity (MBIP) [1,8–10]. The mode of deformation significantly influences the level of work hardening and plays a major role in determining the mechanical properties [7]. One of those material characteristics is the material's yield strength (YS). In the past, various attempts have been made to model the YS of HMnS based on the chemical composition using linear regression [5,11–13]. However, these models often imply negative solid-solution strengthening by some of the alloying elements, like Mn, that are usually attributed to positive strengthening effects [2,14,15].

Interactions of carbon-metal pairs [16–18] and clusters are described by the concept of short-range ordering (SRO) phenomena [19] in HMnS and might offer an explanation for an additional increase in strength that might lead to a defective linear regression when using standard methods to calculate the impact of solid solution strengthening [14,20,21]. Short-range ordered clusters are a result of local differences in ordering energies that make particular unit cell configurations preferential to inhibit a carbon atom [22]. Ab initio calculations employing density-functional theory (DFT) have been shown to be suitable for calculations of ordering energies in HMnS for several chemical compositions. For example, thermodynamic predictions of κ-phase stability [23], interface energies [24], stacking fault energies [25], and hydrogen trap sites [26] showed good agreement with the experimental data.

Cell models [27] have been used to determine the location of carbon atoms in relation to the matrix or substitutional alloying atoms. Other studies show that the interaction of carbon atoms with each other is very repulsive and, thus, influences the diffusion behavior [4,25,28–30], which also points to a coordination of carbon atoms in octahedral sites dependent of surrounding solute atoms. Additionally, the position of C also influences the stacking fault energy of the material [31].

Short-range ordering has also been used to explain the pronounced planar glide that HMnS exhibit because of glide plane softening that occurs when short-range ordered C-Mn clusters are destroyed by dislocation stresses reorienting the carbon atom from its octahedral position to a tetrahedral position. This phenomenon also explains the serrated flow curves that can be found during tensile testing of HMnS, as the back stresses of dislocation pile-ups lead to an increase in strength. The reorientation of the carbon atom then leads to a temporary destruction of the short-range ordered cluster and as a result, a rapid drop in strength because of subsequent softening of this glide plane [32]. However, short-range ordering can be restored by short-range jumps of C back from the tetrahedral site to its initial octahedral site [1,2]. Those short-range jumps might be activated by temperature, but already occur at low temperatures where diffusion processes are usually negligible [1,32].

In this study, a basic cell model for the nearest-neighboring atoms of a central octahedral carbon atom is expanded to include the corner atoms of an elementary cell of an austenitic face-centered cubic (FCC) unit cell, which can be described as the second-nearest neighbors of the central octahedral site in an austenitic cell [27]. Results of ab initio calculations are employed to show that the position of Mn and Al being coordinated as a nearest neighbor or second-nearest neighbor to a central carbon atom is of major importance to the energy level of the system [33]. It is assumed that this also holds true in the case of a central interstitial nitrogen atom [2,33,34] but, in this work, the authors will focus solely on carbon. Based on the calculated ordering energies, the yield strength of HMnS is a result of a combination of short-range ordered clusters, regular solid-solution hardening, grain size effects, and the base strength of an austenitic lattice [14,35].

2. Materials and Methods

Many models for the calculation of the yield strength of austenitic steels have been published in the past. Those models generally employ two mechanisms to describe the yield strength: a strengthening by grain size reduction, the Hall-Petch effect, and the alloying elements' contribution to the material's strength [15,36,37]. Many of those models have used databases with significant amounts of different alloys to employ a linear regression in order to get parameters that determine the effect of alloying elements on the yield strength of the austenitic steel. Assuming that SRO in HMnS influences the yield strength σ_Y significantly, a term describing this effect has to be added to the equation [14], resulting in:

$$\sigma_Y = \sigma_0 + \sigma_{SS} + \sigma_{GS} + \sigma_{SRO} \tag{1}$$

where σ_0 is the base strength of a pure austenitic iron lattice and is assumed to be 90 MPa [14]. σ_{SS} is the contribution of alloying elements to the change in yield strength relative to a pure austenitic lattice. Since this paper focuses on HMnS with the main alloying elements C, Mn, and Al, those elements are assumed to be the main contributors to solid solution strengthening and the contributions of other alloying elements are neglected. It is assumed that this does not lead to a large mistake in calculating σ_{SS} because, apart from Mn and Al, no substitutional alloying element exceeds alloying contents of more than 1 wt. %.

σ_{SRO} is the change in yield strength stemming from the existence of short-range ordered regions in the material. Based on the work of [14], this contribution to yield strength can be calculated as:

$$\sigma_{SRO} = M\tau_{SRO} = M\left(\frac{E_{Order}}{b^3}\right) = M\left(\frac{E_{random} - E_{SRO}}{b^3}\right) \tag{2}$$

where $M = 3.06$ is the Taylor-factor for fcc with $b = \frac{\alpha_\gamma}{\sqrt{2}}$ being the magnitude of the Burgers vector and the lattice parameter in austenite in Å depending on the molar fractions x_i of the alloying elements and temperature in K [38]:

$$\alpha_\gamma = \left(\begin{array}{c} 3.5780 + 0.033x_C + 0.00095x_{Mn} - 0.0002x_{Ni} \\ + 0.0006x_{Cr} + 0.0056x_{Al} + 0.0031x_{Mo} + 0.0018x_V \end{array} \right) \times \left(1 + 2.065 \times 10^{-9} \times (T - 300) \right) \quad (3)$$

Looking at Equation (3), the temperature dependence of the lattice parameter is assumed to be linear.

σ_{GS} describes the strengthening by grain refinement in accordance with the Hall-Petch effect and can be formulated as:

$$\sigma_{CS} = k_y d^{-1/2} \quad (4)$$

Even though k_y depends on the cooling procedure and the carbon content [39], the value $k_y = 11.3 \, \text{MPa mm}^{1/2}$ seems to be a reasonable assumption for HMnS in general, based on the findings in the literature [1,2,14,39–41].

As can be seen in Figure 1, one unit cell consists of 14 atoms situated on the lattice positions of the FCC bracket and one central octahedral interstitial site, where a single carbon atom can be situated.

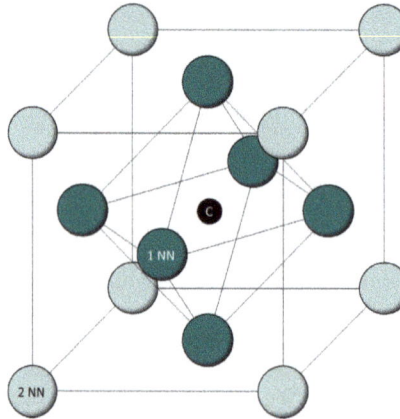

Figure 1. FCC-unit cell with 14 lattice positions for matrix atoms and one central interstitial atom.

First-nearest neighbors (1NN) are the six direct neighbors of the carbon atom. In Figure 1, they are shown in green. Second-nearest neighbors (2NN) in this model are the grey atoms on the corners of the FCC lattice. Although third-nearest neighbors (3NN) can be included in first-principle calculations, a recent study has shown that neither 3NN Al atoms nor 3NN Mn atoms have a positive impact o the ordering energy [33]. To reduce the complexity of the calculations and limit the amount of possible cell configurations, the presence of 3NN alloying atoms is not accounted for in the analysis carried out in this paper. Instead, it is assumed that the material is the sum of FCC-unit cells that do not impact each other. This is obviously not true in reality, but for the purposes of this study, which is aimed at providing a basic model to connect ab initio calculations with experimental results, it is suitable, since it leads to reduced mathematical complexities. The term "cell configuration" is used to describe the combination of alloying atoms and matrix atoms in the cell. "Coordination sphere" is used throughout this paper to refer to the position of atoms relative to the central carbon atom.

The configurations of the cells based on the alloying contents for this 14-cell model are calculated according to [27]:

$$n_{kl} = \frac{14!}{(14 - k - l)!k!l!} \theta_{Fe}^{(14 - k - l)} \theta_{Mn}^k \theta_{Al}^l \quad (5)$$

where n_{kl} is the fraction of all cells that have a configuration with k Mn atoms and l Al atoms with θ_i being the molar fraction of Fe, Mn, and Al, denying the existence of other substitutional alloying elements and ignoring C, since carbon is an interstitial alloying element. Since diffusion rates of substitutional atoms in austenitic steels are very low, it is assumed that the distribution of substitutional elements according to Equation (5) is fixed and that the carbon atom is distributed in accordance with a minimum ordering energy. For clusters of nitrogen and chromium, this approach has been shown to be a realistic approximation, since the presence of nitrogen does not negatively influence the homogeneous distribution of chromium [34]. Cells of type n_{kl} that contain a carbon atom are classified as n_{Ckl}. It should be noted that the assumption that substitutional atoms are distributed according to minimum ordering energies with respect to the presence of a central carbon atom is a very strong assumption, which is unlikely to be true in reality. However, without this assumption, the calculation of ordering energies ΔE_{Ckl} is not possible since ab initio results are presently not available for those unfavorable configurations.

The ordering energy ΔE_{Ckl} of a cell with k Mn atoms, l Al atoms and a central carbon atom is the combination of two energy contributions in comparison to a FCC-unit cell consisting only of Fe atoms and a central interstitial carbon atom. First, the energy required to bring a carbon atom in solution ΔE_{Sol} in a FCC-unit cell with respect to the presence of k Mn atoms and l Al atoms in the first two coordination spheres. The second factor is the change in energy level ΔE_{Lat} resulting from a lattice distortion due to the presence of Mn and Al:

$$\Delta E_{Ckl} = \Delta E_{Sol} + \Delta E_{Lat} \tag{6}$$

Ab initio calculations show that Al is preferentially situated in the second coordination sphere and Mn has a positive effect on the ordering energy level if it sits on a lattice position as a 1NN [22,25,33]. According to those studies, this means that the difference in ordering energy is so profound that the likelihood of a cell configuration that contains 1NN Al atoms instead of 2NN Al atoms or 2NN Mn atoms instead of 1NN Mn atoms is very low. Accordingly, for further calculations in this work and to reduce the model's complexity, it is assumed that for any number of Mn atoms up to six, they will always be 1NN and for any number of Al atoms up to 8, they will always be 2NN.

Using the calculated ordering energies ΔE_{Ckl} from [33], the six cells with the lowest ordering energies are listed in Table 1 and shown in Figure 2.

Table 1. Cell configurations of lowest ordering energy ΔE_{Ckl}.

Cell Name n_{Ckl}	k Mn Atoms	l Al Atoms	ΔE_{Ckl} [eV]
n_{C08}	0	8	−1.2867
n_{C18}	1	8	−1.1524
n_{C28}	2	8	−1.0559
n_{C07}	0	7	−0.9304
n_{C68}	6	8	−0.9287
n_{C38}	3	8	−0.9233

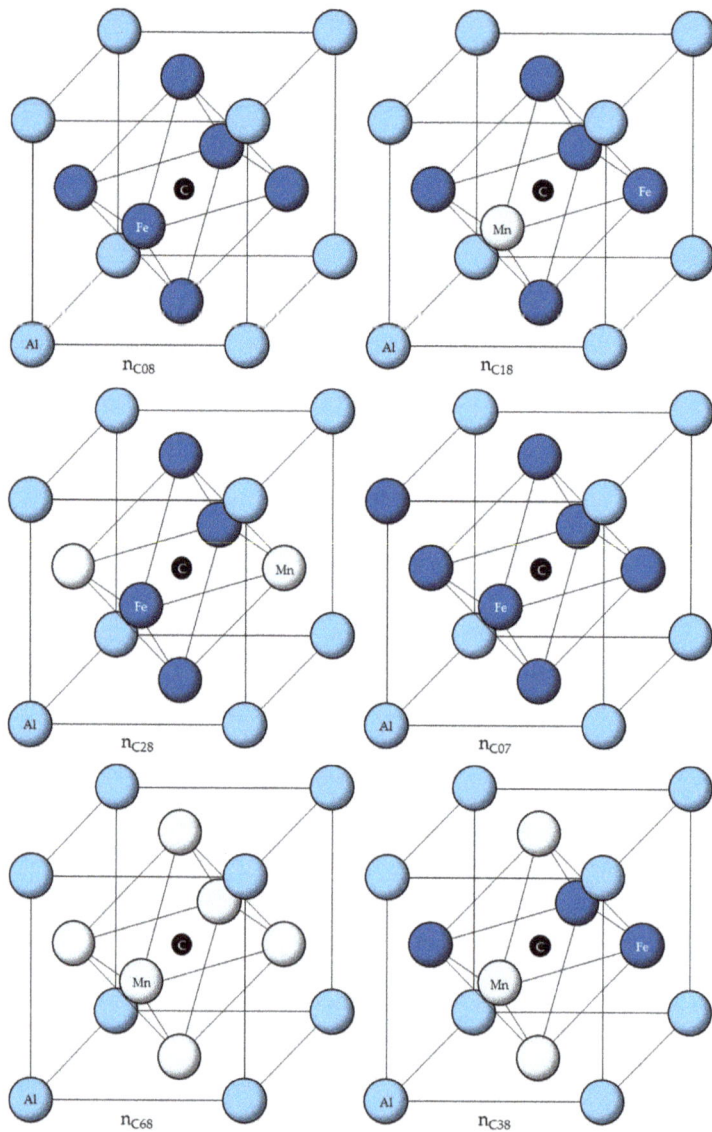

Figure 2. Energetically most favorable cell configurations.

In Equation (2), E_{random} describes the energy state of an unordered alloying system. In this case, a single carbon atom is situated in the central octahedral sites of a randomly chosen FCC-unit cell. This distribution is based on the assumption that there are no preferential octahedral sites, so neither the presence of Mn nor Al influences the position of carbon atoms. Thus, it can be assumed that, for the calculation of E_{random}, the likelihood of the presence of carbon is the same for each type of cell. Using the random C distribution for E_{random}, E_{random} is calculated according to:

$$E_{random} = \sum_{k=0}^{14} \sum_{l=0}^{14} n_{Ckl,random} \Delta E_{Ckl}, \text{ for } k + l \leq 14 \text{ and } k, l \geq 0 \tag{7}$$

$$n_{Ckl,random} = n_{kl} \times \theta_C \qquad (8)$$

Since $\sum n_{kl} = 1$, this method of determining $n_{Ckl,random}$ ensures that carbon is distributed homogeneously in the material, as well as ensuring that every available carbon atom is situated in an octahedral site of a cell.

Contrary to the homogeneous distribution of carbon for E_{random}, the distribution of C for the calculation of E_{SRO} is based on the calculated ordering energies [33]. Since the differences in ordering energy between different cell configurations are significant, as is shown in Table 1, it is assumed that carbon is always situated in the most favorable cell that has not already been assigned a carbon atom. Assuming that at higher temperatures, for example during recrystallization annealing, the diffusion of carbon as an interstitial alloying element is possible, carbon will diffuse to positions with minimal ordering energy in order to minimize the system's Gibbs energy. This means that the cells selected for saturation with carbon are the cells with the lowest ΔE_{Ckl} until $n_{Ckl,SRO} = \theta_C$.

The calculation of ordering energies in the model presented within this work was performed via ab initio simulations based on quantum-mechanical structure optimizations employing density-functional theory (DFT) within the Vienna ab initio Simulation Package (VASP) [33]. The quantum-mechanical structure optimizations are performed based on calculations of lowest-energy states of the electron structure and thus, are conducted at 0 K. Results from ab initio approaches employing DFT have offered good results that are in line with experimental evidence. For example, Mössbauer spectroscopy [17], atom-probe tomography (APT) [42], in situ synchrotron X-ray diffraction measurements [23], and correlative TEM/APT approaches [43] have shown that results from ab initio simulations employing DFT are applicable to systems containing nano-sized κ-carbides that are usually formed during isothermal annealing at temperatures of 600 °C and above. Generally speaking, the effect of temperature changes on the lattice parameter and on intrinsic material properties that derive from electron effects like electrical conductivity is often approximated to be linear [44]. The serrated flow curves of many high-manganese TWIP steels usually occur at temperatures of up to about 200 °C. At higher temperatures, this dynamic strain aging-like effect is either very small or is not detected at all [2,7].

In order to model the temperature dependence of σ_{SRO}, the ab initio results are fitted using a linear approximation of a decrease in σ_{SRO} between 0 K and 500 K. For this purpose, it is assumed that the fact that no serrated flow curve is expected at 500 K can be interpreted as a result of no significant strength increase by short-range ordered clusters.

Based on those assumptions and equations, $\sigma_{Y,calc}$ in MPa was calculated using Equations (2), (3), and (7) for 13 different HMnS with known average grain sizes d_m in μm, listed in Table 2. The alloying contents are given in wt. %.

Table 2. Chemical composition of HMnS used in this study in wt. %; grain size d_m in μm; experimentally determined and calculated yield strength $\sigma_{Y,exp}$ and $\sigma_{Y,calc}$ in MPa, experimental data of alloys marked with * are taken from [14].

Alloy	C	Mn	Al	d_m	$\sigma_{Y,exp}$	$\sigma_{Y,calc}$
X60MnAl17-1	0.56	17.00	1.35	8	328	356
X50MnAl15-1 *	0.49	15.40	1.30	27	294	281
X30MnAl22-1	0.33	22.46	1.21	18	277	275
X30MnAl17-1	0.32	16.80	1.47	17	255	276
X70Mn24	0.71	23.50	0.01	16	334	334
X60Mn23	0.57	23.21	0.01	10	328	334
X50Mn30 *	0.50	30.00	0.00	30	276	273
X50Mn22	0.54	21.95	0.00	14	328	311
X50Mn18 *	0.53	17.90	0.01	8	356	341
X30Mn28	0.28	28.00	0.00	9	291	294
X30Mn23	0.32	22.79	0.01	10	284	291
X30Mn22 *	0.31	22.28	0.00	3	381	384
X30Mn13 *	0.30	12.74	0.01	10	266	274

Experimental yield strength data was taken from the literature [5,14]. Additionally, quasistatic tensile tests have been carried out using a Zwick Z250 tensile testing machine of the AllroundLine by Zwick Roell (Ulm, Germany).

σ_{SS} has been calculated using a linear approximation to explain the differences between experimental and calculated values for the yield strength using the following equation that is a direct result of Equation (1):

$$\sigma_{SS,exp} = \sigma_{Y,exp} - \sigma_{\gamma-Fe} - \sigma_{GS} - \sigma_{SRO} \tag{9}$$

This way, $\sigma_{SS,exp}$ was determined for all steels in Table 2. A regression analysis was performed with respect to the alloying contents of C, Mn, and Al.

The regression analysis lead to the following equation to calculate σ_{SS} with respect to the alloying contents χ_i in weight percent:

$$\sigma_{SS} = 84.55\chi_C + 1.35\chi_{Mn} + 0.44\chi_{Al} \tag{10}$$

Overall, the lattice distortion effect of carbon is by far the strongest contribution to solid solution hardening. This is in accordance with calculations of the misfit of C in an octahedral site using the crystallographic correlation $\frac{r_{OL}}{r_{Fe}} = 0.41$. Comparing the size of the octahedral site to the size of a carbon atom shows that a carbon atom is approximately 8% bigger than the octahedral site, leading to a lattice distortion. Using the lattice parameter calculated according to Equation (3), this size difference can be calculated to approximately 0.03 nm. Comparing this value with the lattice distortion effects by Al and Mn in an austenitic steel according to [45], which are 0.000065 $\frac{nm}{wt. \% Mn}$ and 0.00095 $\frac{nm}{wt. \% Al}$ respectively, the difference in contributions between C and Mn and Al in Equation (10) appear to be reasonable.

3. Results and Discussion

Figure 3 shows that the calculations for the yield strength $\sigma_{Y,calc}$ using the proposed model show good agreement with the experimental data.

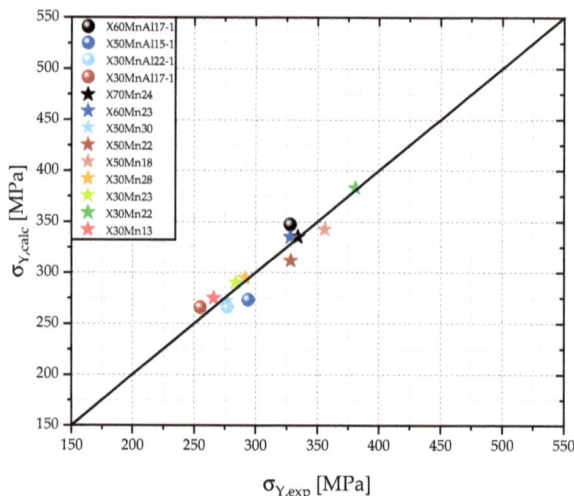

Figure 3. Comparison of calculated and experimental values for σ_Y.

The calculations used in this work lead to differences between $\sigma_{Y,calc}$ and $\sigma_{Y,exp}$ that are smaller than 10%. Those small differences between $\sigma_{Y,calc}$ and $\sigma_{Y,exp}$ might be a result of a negligence of other alloying elements that participate in solid solution hardening, minor inaccuracies of the measurement of the alloying contents, or d_m.

Since σ_{SRO} can also be understood as a deviation from linear solid solution hardening behavior, a close relationship between σ_{SS} and σ_{SRO} seems logical. Higher alloying contents of C, Mn, and Al lead to a higher solid solution strengthening, but also to more carbon atoms integrated in Mn-C, Al-C, and Mn-Al-C clusters, as well as a higher amount of cells with more Al atoms in the second coordination sphere or Mn atoms in the first coordination sphere according to Equation (5).

The individual contributions of solid solution hardening σ_{SS} and σ_{SRO} are illustrated in Figure 4.

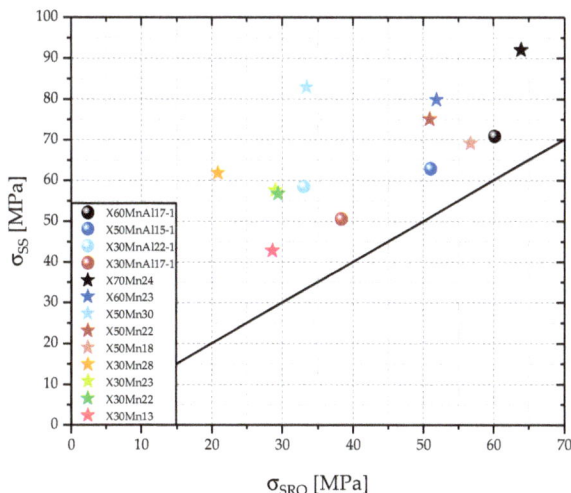

Figure 4. Contributions of σ_{SS} and σ_{SRO} to yield strength.

The strengthening effect of short-range ordered clusters σ_{SRO} is dependent on the chemical composition. For the steel compositions analyzed within the scope of this work, σ_{SRO} at room temperature is approximately in the range of 20 MPa for X30Mn28, containing 0.28 wt. % C and 28.00 wt. % Mn, and 65 MPa for X60MnAl17-1, containing 0.56 wt. % C, 17.00 wt. % Mn, and 1.35 wt. % Al.

Based on Figure 4, one could think, that the connection between σ_{SS} and σ_{SRO} is approximately linear. However, it needs to be kept in mind that the calculation of σ_{SS} is based on linear approximations as well. Additionally, it should be pointed out that although the contribution of σ_{SS} to σ_Y is always greater than that of σ_{SRO}, the proportion between σ_{SS} and σ_{SRO} does not seem to follow a linear trend based on chemical composition. This is illustrated by the line in Figure 4 that represents a relation of $\frac{\sigma_{SS}}{\sigma_{SRO}} = 1$. To further point out the differences in the contributions to σ_Y, the individual contributions of σ_{GS}, σ_{SS}, and σ_{SRO} are portrayed in Figure 5.

According to Figure 5, the contribution of σ_{SRO} to σ_Y is significant. The proportional contribution of σ_{SRO} to σ_Y is in the range of 7.1% for X30Mn28 up to 19.1% for X70Mn24. Since a grain size effect on σ_{SRO} is not assumed in this work, the impact of σ_{SRO} for a given steel composition with larger grain size is higher than for the same steel composition with finer grains. Correspondingly, the proportional contribution of σ_{SRO} that is portrayed in Figure 5 is higher for compositions with larger grains and thus, with smaller grain size strengthening effects σ_{GS}.

The comparison of σ_{SS} and σ_{SRO} for the steels X30Mn13 and X30Mn22 could lead to the conclusion that an increase in Mn content does not significantly influence σ_{SRO}. On the other hand, by comparing X30Mn22 with X30Mn28, it does appear that Mn actually reduces σ_{SRO}. Accordingly, based on these calculations, no clear tendency of the impact of a single alloying element on σ_{SRO} can be found.

For a more detailed analysis, the impact of C and Mn on σ_{SRO} is portrayed in Figures 6 and 7 with regards to the development of σ_{SRO}, σ_{SS}, E_{random}, and E_{SRO} for steels with and without Al.

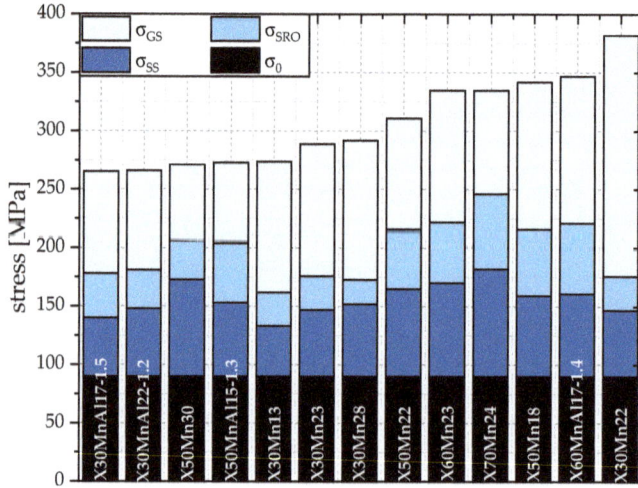

Figure 5. Proportional contribution of σ_{SS}, σ_{SRO}, and σ_{GS} to σ_Y.

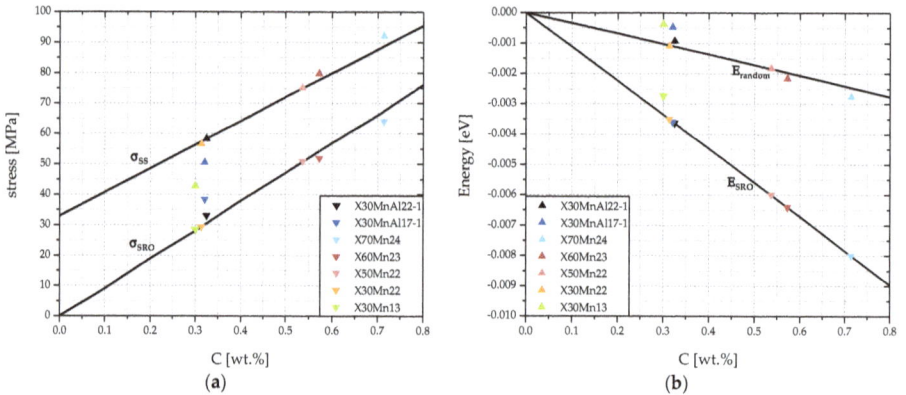

Figure 6. Impact of C on (**a**) σ_{SRO}, σ_{SS}; and (**b**) E_{random} and E_{SRO}.

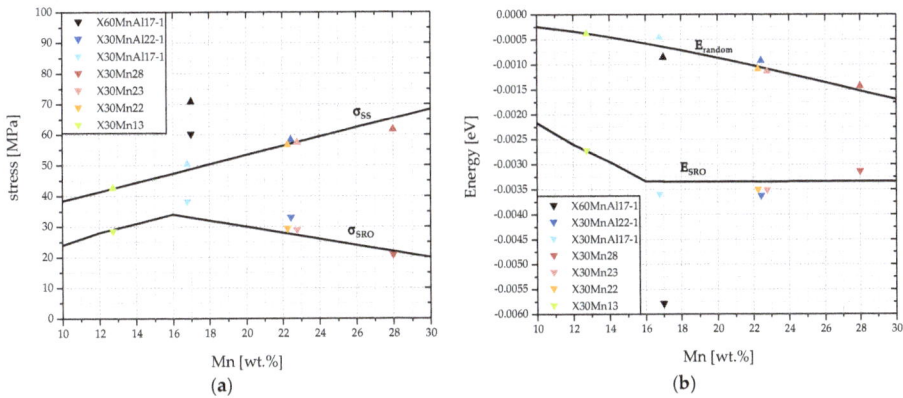

Figure 7. Impact of Mn on (a) σ_{SRO}, σ_{SS}; and (b) E_{random} and E_{SRO}.

Figure 6 analyzes a steel with a constant Mn content of 22 wt. % with increasing C content from 0.0 wt. % up to 0.8 wt. %. Additionally, seven steels that were studied in this work were selected to illustrate the impact of changes in Mn content and Al content.

According to Figure 6a, σ_{SRO} and σ_{SS} increase approximately linearly with C content. This can be explained by an increasing number of cells with carbon atoms. In theory, this increase in σ_{SRO} will only occur as long as there are cells with a preferential structure available, which leads to a decrease in ordering energy E_{SRO}, as can be seen in Figure 6b. If no preferential cells were available, C atoms would have to be situated in non-preferential cells with $\Delta E_{Ckl} > 0$ eV. To a small degree, this effect is included in the calculation of E_{random} because C is assumed to be homogeneously distributed among all cells, including those with positive and, thus, unfavorable values of ΔE_{Ckl}. This explains the lower rate of decrease of E_{random} in Figure 6b compared to E_{SRO}. However, σ_{SRO} will always remain positive because, according to Equation (2), as long as E_{random} is bigger than E_{SRO}, σ_{SRO} will be positive. This premise is always fulfilled by using the assumption in this work, that carbon will be situated in the cells of lowest possible ordering energy first and that the distribution of alloying elements for the calculation of E_{random} and E_{SRO} is the same.

Looking at Figure 6a, the steels with Al, X30MnAl22-1 and X30MnAl17-1, show positive deviations for the calculated values of σ_{SRO} and X30MnAl17-1 shows a negative deviation for σ_{SS}. In the case of solid solution strengthening for X30MnAl17-1, this can be explained by the low Mn content of 16.80 wt. % compared to the 22.00 wt. % used as a basis of calculation of σ_{SS} in Figure 6. Accordingly, X70Mn24 and X60Mn23 show slightly higher values for than σ_{SS} it would be the case for 22.00 wt. % Mn.

The increase of σ_{SRO} can be explained by the impact of the introduction of new cell configurations containing Al atoms. According to Table 1, n_{C08} is the lowest possible energy configuration. However, since the Al content of the steels in this study is very low, only very few cells with the configuration n_{C08} are available. The majority of Al atoms is in cells also containing Mn atoms. According to [33], the introduction of Al atoms as a 2NN into cells with Mn atoms as 1NN does not necessarily lead to a lower ordering energy, but is rather dependent of the number of 1NN Mn atoms. The molar fraction of Mn in a steel containing 22 wt. % Mn obviously depends on the wt. % of the other alloying elements. For the X30MnAl22-1, the molar fraction of Mn is approximately 0.23. According to Equation (5), the most common cell configuration is expected to be n_{30}. The ordering energy of the cell configuration n_{C30} is $\Delta E_{C30} = -0.0501$ eV. This means that this cell configuration is preferential to a pure iron FCC-unit cell. Increasing the number of Al atoms in this cell configuration to 1 and 2 respectively results in $\Delta E_{C31} = -0.0082$ eV and $\Delta E_{C32} = 0.0338$ eV. Both cell configurations are not preferential with respect to n_{C30} which according to previous assumptions means that the cells n_{C31} and

n_{C32} will not contain a carbon atom as long as there are other more preferential cell configurations like n_{30} that do not already contain a carbon atom. In turn, E_{SRO} will not be negatively influenced by cell configurations like n_{C31} and n_{C32}. However, E_{random} will include the negative impact of unfavorable cell configurations which will lead to an increase of E_{random} and a total increase of σ_{SRO}.

Overall, it can be said that the positive impact of Al on σ_{SRO} is a result of an introduction of favorable cell configurations containing Al atoms that inhibit a carbon atom and the avoidance of cell configurations of unfavorable cell configurations.

Figure 7 analyzes a steel with a constant C content of 0.3 wt. % with increasing Mn content from 10 wt. % up to 30 wt. %. Seven steels with varying Al and C content were added to the diagram.

Contrary to the impact of C that is documented in Figure 6, a local maximum of σ_{SRO} at approximately 16 wt. % Mn is observed. Looking at Figure 7b, this can be explained by the stagnation of E_{SRO} for Mn contents of 16 wt. % and higher. At the same time, E_{random} is decreasing, leading to a smaller difference between E_{random} and E_{SRO} and accordingly to a lower σ_{SRO}.

The stagnation of E_{SRO} can be explained by considering the method of calculation for σ_{SRO}. For Al-free steels, the energetically optimal cell is n_{60}. As long as $\theta_C < n_{C60}$, an increase in the amount of C atoms will decrease E_{SRO}. This is due to the underlying assumption that C atoms will be situated in preferential cells of lowest possible ordering energy. At $\theta_C = n_{C60}$, all C atoms are situated in cells of optimal configuration and thus, E_{SRO} is minimized. Due to the way E_{random} is calculated in Equation (7), E_{random} continues to decrease at $\theta_C > n_{C60}$. As a result, σ_{SRO} decreases.

In accordance with Figure 6a, Figure 7a shows that additions of Al and C increase σ_{SRO} significantly.

The contribution of solid solution hardening σ_{SS} increases linearly with the Mn and C content, as it is to be expected considering Equation (10).

To conclude, unlike for the addition of C, Mn additions do not always increase σ_{SRO}. This is a result of a stagnation of E_{SRO} because of a saturation of preferential cells with carbon. In theory, the same effect should be noticeable for C and Al, as well. However, because of the comparably low alloying contents of those elements, it is very unlikely that a point of saturation will be reached in common alloys.

This decrease in σ_{SRO} can be understood as the impact of the deviation from linear solid solution hardening. If the ratio of Mn atoms to Fe atoms becomes increasingly larger and the amount of cells with randomly-distributed Mn increases, the impact of an actual ordering of Mn decreases since fewer cells are brought into short-range ordering to minimize the ordering energy.

Overall, the change of E_{Order} with varying contents of alloying elements can be understood as the effectiveness of an alloy addition of those elements with respect to the increase in strength by short-range ordering. Since it is assumed that σ_{SS} follows a linear trend and σ_{GS} is fixed, $\left| \frac{\delta E_{Order}}{\delta x_i} \right|$ is an indicator for the effectiveness of an addition of an alloying element. A higher $\left| \frac{\delta E_{Order}}{\delta x_i} \right|$ means a greater efficiency in employing an addition of the alloying element i. Looking at Figures 6 and 7, C and Al both appear to be more efficient than Mn in increasing the yield strength by short-range ordering.

In order to compare the accuracy of the model used in this work to calculate σ_Y with other descriptive models, $\sigma_{Y,model}$ according to models by Pickering [15], Lorenz [15], Bouaziz [11], De Cooman [2], and Choi [46] has been calculated. The equations used for $\sigma_{Y,model}$ are listed in Table 3. The coefficients used for the alloying contents are in $\frac{MPa}{wt. \%}$, coefficients for grain size influence are in MPam$^{0.5}$ unless stated otherwise. It needs to be pointed out that although all of the models are for austenitic steels, only the models of Bouaziz, De Cooman, and Choi are specifically for HMnS. Additionally, some of the models include coefficients for the Hall-Petch effect while others neglect them. In order to be able to compare the models, the grain size effect has been calculated according to Equation (4) and added to the model calculations of Choi and Bouaziz.

Table 3. Solid solution hardening models in the literature.

Model	$\sigma_{Y,model}$
Pickering	$68 + 493N + 354C + 3.7Cr + 14Mo + 20Si + 0.22d^{-1/2}$
Lorenz	$40 + 450N + 525C + 2Mn + 8.4Cr + 22Mo + 5Cu + 0.77d^{-1/2}$
Bouaziz	$228 + 137C - 2Mn + \sigma_{GS}$
De Cooman	$189 + 413/\sqrt{(d[\mu m])}$
Choi	$97 + 279C - 1.5Mn + 49.6Si + 20.5Al + \sigma_{GS}$

Figure 8 compares $\sigma_{Y,model}$ with $\sigma_{Y,calc}$ and $\sigma_{Y,exp}$.

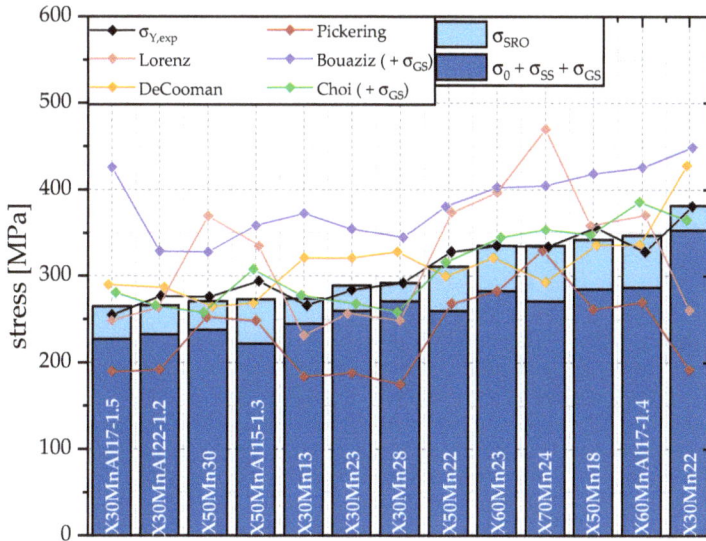

Figure 8. Comparison of models of σ_Y for the steels used in this paper.

Due to the difference in the steel concepts those models were developed for, significant deviations from $\sigma_{Y,exp}$ can be seen. Those deviations are a result of a focus on other alloying elements. For example, Pickering does not consider the impact of Mn on yield strength. Nevertheless, the models of Choi [46] and De Cooman [2] offer reasonable results, but are not as close to the experimental data as the model developed in this work. Additionally, the model by De Cooman does not take compositional variances into account, while the model of Choi considers the impact of Mn on yield strength to be negative.

The models of Pickering and Lorenz are not applicable to HMnS based on the results of the calculations. The model by Bouaziz tends to overestimate the yield strength.

4. Conclusions

A basic ab initio-based model for the calculation of σ_Y has been developed. The model shows good agreement with the experimental values. Discrepancies between experimental $\sigma_{Y,exp}$ and calculation results $\sigma_{Y,calc}$ are smaller than 10%. While many complexity-reducing assumptions have been made that lead to inaccuracies when calculating σ_{SRO}, this model portrays that the general concept of combining ab initio simulations with experimental data can lead to feasible results on a macroscopic scale. This is also shown by the calculated positive impact of Mn and Al on yield strength by solid solution strengthening in a reasonable range.

The impact of the alloying content of C, Mn, and Al on the strengthening by short-range ordering was analyzed. For the given steel compositions with fully austenitic microstructures in this work, the values for σ_{SRO} are as high as approximately 65 MPa. This significant strengthening leads to a proportional contribution to σ_Y of nearly 20%.

While an increase in C and Al increases σ_{SRO} as long as a sufficient number of atoms of substitutional alloying elements are available to form preferential cell structures, a limiting factor for the effectiveness of strength increases by Mn was found.

The impact of short-range ordered clusters during deformation needs to be evaluated. A higher amount of more stable, energetically favorable clusters might not only lead to an increase in σ_{SRO} and σ_Y, it might also impact the beginning of serrations in the flow curve, as well as their amplitude if the serrations are resulting from short-distance carbon atom jumps associated with the cutting of clusters.

Acknowledgments: The authors are thankful for the financial support from Deutsche Forschungsgemeinschaft (DFG) within the Project C2: "Material Properties of High Mn-Steels" of the Collaborative Research Centre (SFB) 761 "Steel-ab initio". The provision of experimental data by Sebastian Wesselmecking (Steel Institute, RWTH Aachen University) is very much appreciated as well as the provision of ab initio calculation results by Dronskowski, Tobias Timmerscheidt, and Dmitri Bogdanovski (Department of Inorganic Chemistry, RWTH Aachen University) of Project A1: "ab initio Quantum chemistry of the Fe-Mn-C system" of SFB 761 "Steel-ab initio".

Author Contributions: Simon Sevsek analyzed the data. Simon Sevsek and Wolfgang Bleck wrote the paper.

Conflicts of Interest: The authors declare no conflict of interest.

References

1. De Cooman, B.C.; Kwon, O.; Chin, K.-G. State-of-the-knowledge on TWIP steel. *Mater. Sci. Technol.* **2012**, *28*, 513–527. [CrossRef]

2. De Cooman, B.C. High Mn TWIP steel and medium Mn steel. In *Automotive Steels: Design, Metallurgy, Processing and Applications*, 1st ed.; Rana, R., Singh, S.B., Eds.; Woodhead Publishing: Cambridge, UK, 2017; pp. 317–385. ISBN 9780081006382.

3. Haase, C.; Ingendahl, T.; Güvenc, O.; Bambach, M.; Bleck, W.; Molodov, D.A.; Barrales-Mora, L.A. On the applicability of recovery-annealed Twinning-Induced Plasticity steels: Potential and limitations. *Mater. Sci. Eng. A* **2016**, *649*, 74–84. [CrossRef]

4. Abbasi, A.; Dick, A.; Hickel, T.; Neugebauer, J. First-principles investigation of the effect of carbon on the stacking fault energy of Fe–C alloys. *Acta Mater.* **2011**, *59*, 3041–3048. [CrossRef]

5. Bouaziz, O.; Zurob, H.; Chehab, B.; Embury, J.D.; Allain, S.; Huang, M. Effect of chemical composition on work hardening of Fe–Mn–C TWIP steels. *Mater. Sci. Technol.* **2011**, *27*, 707–709. [CrossRef]

6. Tewary, N.K.; Ghosh, S.K.; Chatterjee, S. Effect of Al Content in Low Carbon High Manganese TWIP Steel. *Key Eng. Mater.* **2016**, *706*, 16–22. [CrossRef]

7. Bäumer, A.; Jimenez, J.A.; Bleck, W. Effect of temperature and strain rate on strain hardening and deformation mechanisms of high manganese austenitic steels. *Int. J. Mater. Res.* **2010**, *101*, 705–714. [CrossRef]

8. Choi, K.; Seo, C.-H.; Lee, H.; Kim, S.K.; Kwak, J.H.; Chin, K.G.; Park, K.-T.; Kim, N.J. Effect of aging on the microstructure and deformation behavior of austenite base lightweight Fe–28Mn–9Al–0.8C steel. *Scr. Mater.* **2010**, *63*, 1028–1031. [CrossRef]

9. Gutierrez-Urrutia, I.; Raabe, D. Microbanding mechanism in an Fe–Mn–C high-Mn twinning-induced plasticity steel. *Scr. Mater.* **2013**, *69*, 53–56. [CrossRef]

10. Liu, F.; Dan, W.J.; Zhang, W.G. Strain hardening model of TWIP steels with manganese content. *Mater. Sci. Eng. A* **2016**, *674*, 178–185. [CrossRef]

11. Bouaziz, O.; Allain, S.; Scott, C.P.; Cugy, P.; Barbier, D. High manganese austenitic twinning induced plasticity steels: A review of the microstructure properties relationships. *Curr. Opin. Solid State Mater. Sci.* **2011**, *15*, 141–168. [CrossRef]

12. Bouaziz, O. Strain-hardening of twinning-induced plasticity steels. *Scr. Mater.* **2012**, *66*, 982–985. [CrossRef]

13. Ohkubo, N.; Miyakusu, K.; Uematsu, Y.; Kimura, H. Effect of Alloying Elements on the Mechanical Properties of the Stable Austenitic Stainless Steel. *ISIJ Int.* **1994**, *34*, 764–772. [CrossRef]

14. Kang, J.-H.; Ingendahl, T.; von Appen, J.; Dronskowski, R.; Bleck, W. Impact of short-range ordering on yield strength of high manganese austenitic steels. *Mater. Sci. Eng. A* **2014**, *614*, 122–128. [CrossRef]

15. Sieurin, H.; Zander, J.; Sandström, R. Modelling solid solution hardening in stainless steels. *Mater. Sci. Eng. A* **2006**, *415*, 66–71. [CrossRef]

16. Oda, K.; Fujimura, H.; Ino, H. Local interactions in carbon-carbon and carbon-M (M: Al, Mn, Ni) atomic paris in FCC gamma-iron. *J. Phys. Condens. Matter* **1994**, *6*, 679–692. [CrossRef]
17. Bentley, A.P. Ordering in Fe–Mn–Al–C austenite. *J. Mater. Sci. Lett.* **1986**, *5*, 907–908. [CrossRef]
18. McLellan, R.B.; Dunn, W.W. A quasi-chemical treatment of interstitial solid solutions: It application to carbon austenite. *J. Phys. Chem. Solids* **1969**, *30*, 2631–2637. [CrossRef]
19. Pelton, A.D.; Kang, Y.-B. Modeling short-range ordering in solutions. *Int. J. Mater. Res.* **2007**, *98*, 907–917. [CrossRef]
20. Honeycombe, R.W.K. The effect of temperature and alloying additions on the deformation of metal crystals. *Progress Mater. Sci.* **1961**, *9*, 95–120. [CrossRef]
21. Oda, K.; Kondo, N.; Shibata, K. X-ray Absorption Fine Structure Analysis of Interstitial (C, N)-Substitutional (Cr) Complexes in Austenitic Stainless Steels. *ISIJ Int.* **1990**, *30*, 625–631. [CrossRef]
22. Von Appen, J.; Dronskowski, R. Carbon-Induced Ordering in Manganese-Rich Austenite—A Density-Functional Total-Energy and Chemical-Bonding Study. *Steel Res. Int.* **2011**, *82*, 101–107. [CrossRef]
23. Song, W.; Zhang, W.; von Appen, J.; Dronskowski, R.; Bleck, W. k-Phase Formation in Fe–Mn–Al–C Austenitic Steels. *Steel Res. Int.* **2015**, *86*, 1161–1169. [CrossRef]
24. Raabe, D.; Roters, F.; Neugebauer, J.; Gutierrez-Urrutia, I.; Hickel, T.; Bleck, W.; Schneider, J.M.; Wittig, J.E.; Mayer, J. Ab initio-guided design of twinning-induced plasticity steels. *MRS Bull.* **2016**, *41*, 320–325. [CrossRef]
25. Medvedeva, N.I.; Park, M.S.; van Aken, D.C.; Medvedeva, J.E. First-principles study of Mn, Al and C distribution and their effect on stacking fault energies in fcc Fe. *J. Alloys Compd.* **2014**, *582*, 475–482. [CrossRef]
26. Timmerscheidt, T.; Dey, P.; Bogdanovski, D.; von Appen, J.; Hickel, T.; Neugebauer, J.; Dronskowski, R. The Role of κ-Carbides as Hydrogen Traps in High-Mn Steels. *Metals* **2017**, *7*, 264. [CrossRef]
27. McLellan, R.B. Cell models for interstitial solid solutions. *Acta Metall.* **1982**, *30*, 317–322. [CrossRef]
28. Bhadeshia, H.K.D.H. Diffusion of carbon in austenite. *Met. Sci.* **1981**, *15*, 477–479. [CrossRef]
29. McLellan, R.B.; Ko, C. The C–C interaction energy in iron-carbon solid solutions. *Acta Metall.* **1987**, *35*, 2151–2156. [CrossRef]
30. McLellan, R.B.; Ko, C. The diffusion of carbon in austenite. *Acta Metall.* **1988**, *36*, 531–537. [CrossRef]
31. Gholizadeh, H.; Draxl, C.; Puschnig, P. The influence of interstitial carbon on the g-surface in austenite. *Acta Metall.* **2013**, *61*, 341–349. [CrossRef]
32. Lee, S.-J.; Kim, J.; Kane, S.N.; DeCooman, B.C. On the origin of dynamic strain aging in twinning-induced plasticity steels. *Acta Metall.* **2011**, *59*, 6809–6819. [CrossRef]
33. Timmerscheidt, T.; Dronskowski, R. An Ab Initio Study of Carbon-Induced Ordering in Austenitic Fe–Mn–Al–C Alloys. *Steel Res. Int.* **2017**, *88*, 1–10. [CrossRef]
34. Sumin, V.V.; Chimid, G.; Rashev, T.; Saryivanov, L. The Neutron-Spectroscopy Proof of the Strong Cr–N Interactions in Nitrogen Stainless Steels. *Mater. Sci. Forum* **1999**, *318–320*, 31–40. [CrossRef]
35. Bouaziz, O.; Barbier, D. Benefits of Recovery and partial Recrystallization of Nano-Twinned Austenitic Steels. *Adv. Eng. Mater.* **2013**, *15*, 976–979. [CrossRef]
36. Fiore, N.F.; Bauer, C.L. Binding of solute atoms to dislocations. *Progress Mater. Sci.* **1968**, *13*, 85–134. [CrossRef]
37. Butt, M.Z.; Feltham, P. Solid-solution hardening. *J. Mater. Sci.* **1993**, *28*, 2557–2576. [CrossRef]
38. Saeed-Akbari, A.; Imlau, J.; Prahl, U.; Bleck, W. Derivation and Variation in Composition-Dependent Stacking Fault Energy Maps Based on Subregular Solution Model in High-Manganese Steels. *Metall. Mater. Trans. A* **2009**, *40*, 3076–3090. [CrossRef]
39. Kang, J.-H.; Duan, S.; Kim, S.-J.; Bleck, W. Grain Boundary Strengthening in High Mn Austenitic Steels. *Metall. Mater. Trans. A* **2016**, *47*, 1918–1921. [CrossRef]
40. Kang, S.; Jung, Y.-S.; Jun, J.-H.; Lee, Y.-K. Effects of recrystallization annealing temperature on carbide precipitation, microstructure and mechanical properties in Fe–18Mn–0.6C–1.5Al TWIP steel. *Mater. Sci. Eng. A* **2010**, *527*, 745–751. [CrossRef]
41. De las Cuevas, F.; Reis, M.; Ferraiuolo, A.; Pratolongo, G.; Karjalainen, L.P.; Alkorta, J.; Gil Sevillano, J. Hall-Petch relationship of a TWIP steel. *Key Eng. Mater.* **2010**, *423*, 147–152. [CrossRef]
42. Bartlett, L.N.; van Aken, D.C.; Medvedeva, J.; Isheim, D.; Medvedeva, N.I.; Song, K. An Atom Probe Study of Kappa Carbide Precipitation and the Effect of Silicon Addition. *Metall. Mater. Trans. A* **2014**, *45*, 2421–2435. [CrossRef]

43. Yao, M.J.; Welsch, E.; Ponge, D.; Haghighat, S.; Sandlöbes, S.; Choi, P.; Herbig, M.; Bleskov, I.; Hickel, T.; Lipinska-Chwalek, M.; et al. Strengthening and strain hardening mechanisms in a precipitation-hardened high-Mn lightweight steel. *Acta Mater.* **2017**, *140*, 258–273. [CrossRef]
44. Gottstein, G. *Physikalische Grundlagen der Materialkunde*, 3rd ed.; Springer: Berlin/Heidelberg, Germany, 2007; ISBN 978-3-540-71104-9.
45. Li, M.; Chang, H.; Kao, P.; Gan, D. The effect of Mn and Al contents on the solvus of κ phase in austenitic Fe–Mn–Al–C alloys. *Mater. Chem. Phys.* **1999**, *59*, 96–99. [CrossRef]
46. Choi, W.S.; de Cooman, B.C.; Sandlöbes, S.; Raabe, D. Size and orientation effects in partial dislocation-mediated deformation of twinning-induced plasticity steel micro-pillars. *Acta Mater.* **2015**, *98*, 391–404. [CrossRef]

![metals logo] *metals*

MDPI

Article

On the Mn–C Short-Range Ordering in a High-Strength High-Ductility Steel: Small Angle Neutron Scattering and Ab Initio Investigation

Wenwen Song [1,*], Dimitri Bogdanovski [2], Ahmet Bahadir Yildiz [3,†], Judith E. Houston [4], Richard Dronskowski [2,5] and Wolfgang Bleck [1]

[1]   Steel Institute, RWTH Aachen University, Intzestraße 1, 52072 Aachen, Germany; bleck@iehk.rwth-aachen.de
[2]   Institute of Inorganic Chemistry, RWTH Aachen University, Landoltweg 1, 52074 Aachen, Germany; dimitri.bogdanovski@ac.rwth-aachen.de (D.B.); drons@HAL9000.ac.rwth-aachen.de (R.D.)
[3]   Department of Materials Science and Engineering, KTH Royal Institute of Technology, Brinellvägen 23, S-100 44 Stockholm, Sweden; abyildiz@kth.se
[4]   Jülich Centre for Neutron Science (JCNS) at Heinz Maier-Leibnitz Zentrum (MLZ), Forschungszentrum Jülich GmbH, Lichtenbergstraße 1, 85748 Garching, Germany; j.houston@fz-juelich.de
[5]   Jülich-Aachen Research Alliance (JARA-HPC), RWTH Aachen University, 52056 Aachen, Germany
*   Correspondence: wenwen.song@iehk.rwth-aachen.de; Tel.: +49-241-80-95815
†   The author did the research work in the Steel Institute at RWTH Aachen University, Germany, and is currently working in the Department of Materials Science and Engineering at KTH Royal Institute of Technology, Sweden.

Received: 29 November 2017; Accepted: 5 January 2018; Published: 10 January 2018

Abstract: The formation of Mn–C short-range ordering (SRO) has a great influence on the mechanical properties of high-Mn steels. In the present work, the formation of Mn–C SRO during recrystallization of an X60Mn18 steel was investigated by means of a combined study employing small angle neutron scattering (SANS) and ab initio ground-state energy calculations based on density-functional theory. The SANS measurements prove the presence of Mn–C SRO in the recrystallization annealed X60Mn18 steel and indicate the evolution of the SRO during recrystallization. The results show that with the increase in annealing time, the mean size of the Mn–C SRO decreases, whereas the number density increases. The ab initio calculations well describe the energetically favored condition of Mn–C SRO and provide the theoretical explanation of the clustering formation and evolution in the X60Mn18 steel. The stress-strain curve of the X60Mn18 steel exhibits a high strain-hardening rate and the plastic deformation is characterized with a series of serrations during a uniaxial tensile test. In the end, the correlation between Mn–C SRO and the serrated flow of high-Mn steels is further discussed.

Keywords: high-Mn steel; short-range ordering; small angle neutron scattering; ab initio calculations; density-functional theory; Portevin-Le Chatelier effect

1. Introduction

High-Mn (15–30 wt %) austenitic steels are well known for their high ultimate tensile strength over 1000 MPa and total elongation over 60% at room temperature [1–4]. Due to the excellent strain hardening of the high-Mn steels, the superior combination of high strength and ductility makes these steel grades a good choice for the automotive industry for the manufacture of complex body geometries, yielding improved crashworthiness and reduced weight at the same time [5–7]. A unique characteristic of high-Mn steels is the significant serration phenomenon (jerky flow) during deformation. These serrations strongly influence the mechanical properties of the materials, for example, they increase the flow stress, the tensile strength and the work hardening rate, etc.

The strain-hardening behavior of high-Mn steels, particularly of the X60Mn18 steel, is mainly explained by the dynamic Hall-Petch effect, dislocation evolution and dynamic strain aging (DSA) [2,4,8–13]. During the deformation of high-Mn steels, the formation of twins occurs with an increasing number of twin boundaries. As in the case of grain boundaries, the twin boundaries behave like obstacles against the movement of dislocations, decrease the dislocation mean free path and promote dislocation accumulation [4,14,15]. The serrated stress-strain curves are the main indicator of the DSA mechanism. It has been suggested that DSA in Fe–Mn–C steels involves the interaction of dislocations [4,16,17] and stacking fault regions [12,13] with point defect complexes that include interstitial C.

The formation and motion of DSA-caused Portevin-Le Chatelier (PLC) deformation bands result in a serrated flow and underline the deformation mechanism [13]. It is widely known that the serration flow is correlated with the presence of Mn–C agglomerations/short range ordering (SRO) and their interactions with dislocations in steels. Although the serration phenomenon and DSA effect have been discussed for a long time, little experimental work on Mn–C SRO has been reported so far. In particular, the mechanism of formation of this type of SRO and its evolution during the annealing process, which is very important for the explanation of the serrated stress-strain behavior of high-Mn steels, is so far unknown. Due to the very fine scale of SRO at the atomic level and the consequent technical difficulties in performing experimental measurements, there is very limited information on the SRO in literature. The characterization of the Mn–C SRO using atom probe tomography (APT) requires complicated sample preparation procedures and comes with a significant statistical error due to the limited probe size [18,19]. Marceau et al. [18] stated the difficulties in analyzing the short-range ordered clusters in Fe–Mn–C steels, such as the correlated field evaporation of interstitial C, detection of C as a molecular ion species, and surface migration of C (directional walk). In contrast, small angle neutron scattering (SANS) makes it possible to characterize the nano-sized SRO/short-range clusters (SRC) in the mm^3 volume range.

In the present work, the formation of Mn–C SRO during recrystallization of the X60Mn18 steel was investigated by means of a combined method of SANS and ab initio ground-state energy calculations. The Mn–C SRO was analyzed by SANS in terms of particle size, size distribution, and volume fraction. In order to investigate the local atomic environment of C and to identify the favorable arrangements on an atomistic level, density-functional theory (DFT) calculations were performed for a model system with a chemical composition close to the investigated steel. The present study validates the assumptions drawn from the generalized ab initio models of the Mn–C SRO in high-Mn steels and compares the local and atom-resolved ordering information extracted from the experiments. In the end, the correlation between Mn–C SRO and the serrated flow of the recrystallized X60Mn18 steel is further discussed.

2. Material and Methods

2.1. Material

The investigated X60Mn18 high-Mn steel was cast into an ingot. Hot forging was subsequently carried out at 1423 K (1150 °C). The forged ingot was then homogenized at 1423 K (1150 °C) for 5 h. Then the material was hot rolled, and cold rolled with a thickness reduction of ≈50%. Table 1 lists the chemical composition of the investigated material. The specimens used for the ex-situ SANS measurements and mechanical testing were cut from the 1.5 mm thick cold rolled sheets. In order to compare the impact of annealing duration on the Mn–C SRO formation, the specimens were recrystallization annealed in a salt bath at 800 °C for 2 min and 30 min, followed by quenching in water to room temperature. The tensile tests were carried out using a screw-driven ZWICK Z100 machine (Zwick GmbH & Co. KG, Ulm, Germany) with a strain rate of $0.001\ s^{-1}$. A mechanical strain gauge with a 30 mm gauge length was used to measure the specimen strain.

Table 1. Chemical composition of the investigated X60Mn18 steel.

Element	C	Si	Mn	P	S	Cr	Ni	Al	Fe
wt %	0.594	0.05	18.40	0.007	0.009	0.02	0.04	0.005	Rest

2.2. Small Angle Neutron Scattering (SANS)

2.2.1. SANS Technique

SANS is a well-established, non-destructive method to obtain structural information about the arrangement of atoms and magnetic moments on a mesoscopic scale, from near atomic (nanometer) to near optical (micrometer) sizes, in condensed matter systems and materials from a wide range of applications. Such arrangements may be macromolecules [20], self-assembled polymeric systems [21], biomolecular aggregates [22], precipitates and SRO in metallurgical materials [23,24] or porosities in geological and construction materials [25]. Neutrons interact with matter via short-range nuclear interactions and, thus, detect the nuclei in a sample rather than the diffuse electron cloud observed by X-rays. Therefore, unlike X-rays, neutrons are able to "see" light atoms in the presence of heavier ones and distinguish neighboring elements more easily. In magnetic samples, neutrons are additionally scattered by magnetic moments associated with the unpaired electron spins (dipoles) and, thus, can observe magnetic structures and correlations when the samples are placed in a saturating magnetic field.

In SANS the "visibility" of a selected constituent in a complex system depends upon the contrast—the squared difference between its scattering length density (magnetic or nuclear) and that of the other components [26]. Thus, the visibility of the particles, P, embedded in a matrix, M, depends on the particles' contrast, given by:

$$\Delta \rho_P^2 = (\rho_P - \rho_M)^2 = \left(\frac{\sum_i^{\text{atoms in } P} b_i^P}{v_P} - \frac{\sum_i^{\text{atoms in } M} b_i^M}{v_M} \right)^2 \tag{1}$$

where b_i^M denotes the scattering length of different atoms in M, b_i^P represents those of the atoms in P, and v_M and v_P are the effective volumes occupied by the objects composed of the atoms in the respective sums.

The contrast in a multicomponent system can typically be estimated in a direct way based on the composition information obtained during sample preparation. On the other hand, the contrast in a complex material can be manipulated by isotopic substitution, without changing the macroscopic properties of the material. This means that matching and variation of contrast of different components of the material can be achieved to enable separate analysis of selected components or regions within complex morphologies.

Finally, SANS has the advantage of being non-destructive and of providing information with high statistical accuracy due to averaging over a macroscopic sample volume. In many cases, the small absorption of neutrons allows for the investigation of thick samples as well as the use of bulky ancillaries for studying the materials under special conditions, such as very low or very high temperature, magnetic field or high pressure.

The information in SANS experiments is contained in the neutron scattering intensity measured as a function of the momentum transfer, q, given by:

$$\vec{q} = \vec{k}_i - \vec{k}_f; \quad q = \frac{4\pi}{\lambda} \sin \theta \tag{2}$$

where $\vec{k}_i$ and $\vec{k}_f$ are the incoming and the outgoing neutron wave vectors, λ is the neutron wavelength and θ is the half scattering angle.

Scattering experiments explore matter in reciprocal space: large q values relate to short distances, while small q values relate to larger structures. To observe the mesoscopic scale, SANS is optimized at small scattering angles using long wavelength (cold) neutrons. The contribution to the scattering intensity at angle θ from some scatterers is given by:

$$I(\theta) = T I_0 \frac{d\sigma}{d\Omega} \Delta\Omega \tag{3}$$

where $I(\theta)$ is the intensity scattered into one detector element, $\Delta\Omega$ the solid angle covered by that element, I_0 the incoming neutron flux, and T the sample transmission. The structural information on the sample is contained in the microscopic differential cross-section $d\sigma/d\Omega$, which is the Fourier transform of the scattering length density distribution $\Delta\rho\left(\vec{r}\right)$.

The scattering data are obtained as a microscopic differential cross-section $d\sigma/d\Omega$ on an absolute scale, in cm^{-1}, after normalization of the macroscopic differential cross-section by the sample volume and a comparison with scatterers of a known absolute cross-section (standard samples).

2.2.2. SANS Experiments

Ex-situ SANS experiments were carried out in the KWS-2 beamline of the Heinz Maier-Leibnitz Centre (MLZ) at the FRM II reactor in Garching, Munich [27]. The instrument is optimized for the exploration of the wide momentum transfer q range between 1×10^{-3} nm^{-1} and 5 nm^{-1} by combining classical pinhole, focusing (with lenses) and time-of-flight (with chopper) methods, while simultaneously providing high neutron intensities with an adjustable resolution [27]. A neutron beam with a wavelength λ of 0.5151 nm and a wavelength spread $\Delta\lambda/\lambda = 0.2$ was used. The scattering vector q ($q = 4\pi \sin\theta/\lambda$, where θ is the half scattering angle) [28] range of 0.085 nm^{-1} to 4.2 nm^{-1} was achieved with two different sample-to-detector distances (2 m and 8 m) and 8 m collimation length. This q range allows the characterization of the scatterer size D ($D \approx 2\pi/q$) which ranges between approximately 1 nm and 70 nm. The total measurement durations were set as 30 min and 60 min at 2 m and 8 m sample-to-detector distances, respectively. These experiment durations enabled the achievement of a total number of scattering counts in excess of 1 million. The transmission values were measured at 20 m collimation and 8 m sample-to-detector distances for 60 s.

2.3. Structural Models and Ab Initio Computational Details

To approximate the experimental composition as closely as possible, $Fe_{88}Mn_{20}C_3$ was chosen as the model system for the ab initio calculations. This structure was modeled as a $3 \times 3 \times 3$ supercell of fcc (face-centered cubic)-Fe (space group $Fm\overline{3}m$), resulting in an fcc-Fe matrix consisting of 108 atoms, with 20 of the Fe-atoms then replaced by Mn. The three C atoms occupy interstitial sites that correspond to the central octahedral void in the original fcc-Fe unit cell, thus increasing the total atomic count to 111. Employing this model, the composition of the system arrives at approximately 79.3 at % Fe, 18.0 at % Mn and 2.7 at % C, corresponding to respective values of approx. 81.2 wt %, 18.2 wt % and 0.6 wt %, reflecting sample composition reasonably well.

Four distinct structural models (further referred to as STM) were considered in respect to the Mn distribution, and they are shown in Figure 1: a "disordered" structure approximating a random Mn distribution (STM 1) and three different "ordered" structures. In the latter case, Mn_6 octahedra coordinate the interstitial sites serving as potential carbon locations, the Mn_6C element being an energetically favorable structural motif [29,30] corresponding to the aforementioned Mn–C SRO. This octahedral arrangement of Mn as the nearest neighbor to C, located in the 1st coordination shell (CS) of the latter, has been shown to be energetically favored in prior studies [30]. On the other hand, the cubic arrangement of Mn as a second-nearest neighbor to C, i.e., in the 2nd CS, is disfavored, while arrangements with Mn located in further shells have no energetic impact. Taking these considerations into account, the models for the three "ordered" STMs were restricted to the favored octahedral 1st CS arrangement. Said ordered STMs differ in the spatial arrangement of Mn atoms: in the first case, four Mn_6 octahedra are located in one plane in the individual corners of the unit cell (STM 2). In the second case, two groups of two octahedral units each are diagonally opposed to each other, with one group comprising Mn_6 octahedra and the other actually consisting of Mn_5Fe octahedra due to stoichiometry constraints (STM 3). In the third case, three corner-sharing octahedra propagate along the [100] direction, with a fourth corner-sharing Mn_6 octahedron located orthogonally to that axis, forming a rough "T" shape (STM 4). Considering translational symmetry, the ordered structures thus model closely packed groups of Mn_6 octahedra (corner-sharing tetramers) with several "layers" of Fe between them (STM 2), more evenly distributed groups of chemically slightly distinct dimers ($2 \times Mn_6$ and $2 \times Mn_5Fe$) (STM 3) and linear chains of Mn_6 octahedra along one crystallographic direction with regular "branches" (STM 4), respectively. It should be noted that these models were selected under the constraint of computational feasibility, based on known data on attractive Mn–C and repulsive C–C interactions [30,31], in which inter-atom distance, and thus distribution, is a key factor. A full sampling of statistically possible configurations requires a multitude of different environments, larger cells and usage of mesoscopic methods, and this is presently outside the scope of an atomistic ab initio approach.

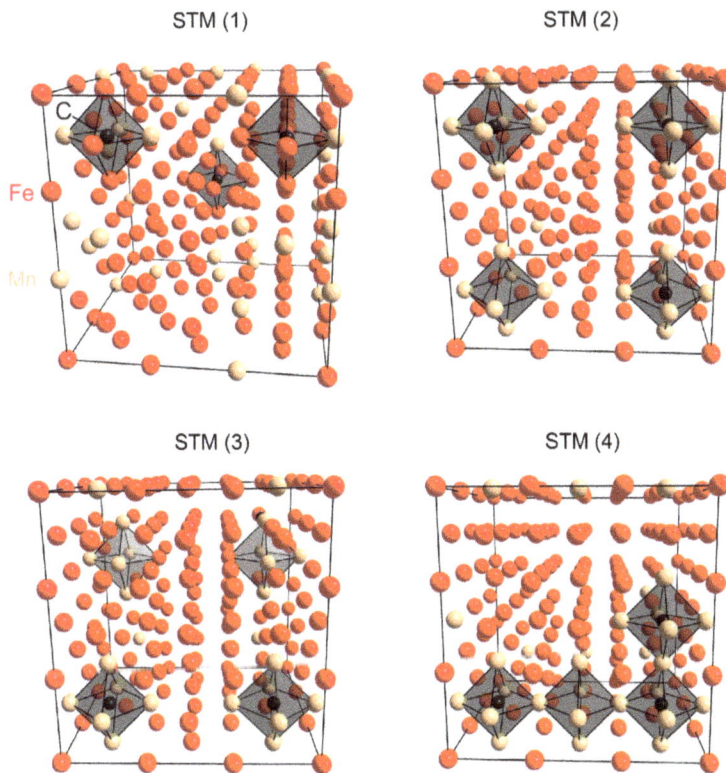

Figure 1. The four different structural models (STM) of the $Fe_{88}Mn_{20}C_3$ system differing in Mn arrangement, each showing one exemplary configuration. Fe atoms are red, Mn atoms beige, C atoms black. Darker grey polyhedra are shown to identify the Mn_6 units in the three ordered systems, with lighter grey polyhedra in structure (3) signifying Mn_5Fe.

Full structural optimizations of all systems were performed via density-functional theory (DFT) calculations using the Vienna ab initio Simulation Package (VASP, version 5.4.1, Computational

Materials Physics, University of Vienna, Vienna, Austria, 2015) [32,33], employing the projector-augmented wave (PAW) method [34,35] for basis set representation, with the energy cut-off set to E_{cut} = 500 eV. Exchange and correlation effects were accounted for using the generalized gradient approximation (GGA) within DFT as parametrized in the well-established functional by Perdew, Burke and Ernzerhof (PBE) [36]. The generation of the k-mesh for partitioning and integration of the Brillouin zone was performed with the Monkhorst-Pack scheme [37], employing a 4 × 4 × 4 grid. Partial occupancies were considered using the Methfessel-Paxton scheme [38] of order 2 with the smearing width set to σ = 0.15 eV. The disordered structural model was generated using the special quasirandom structure (SQS) method [39] as implemented in the Alloy Theoretic Automated Toolkit (ATAT, version 3.36, Axel van de Walle, 2017) package [40,41]. All simulations were performed for a non-spin-polarized case, which approximates a paramagnetic model under standard conditions reasonably well [42]; while approaches explicitly modeling paramagnetism such as the disordered local moment (DLM) model [43] are somewhat more accurate, they are typically much more demanding in terms of computation and thus were not employed here due to the system size involved. Both fcc-Fe and all allotropes of Mn are antiferromagnetic in the ground state, but their respective Néel temperatures lie well below room temperature [44,45]; thus, the non-spin-polarized case is a very good approximation for the purposes of this study. Nevertheless, it must be pointed out that magnetic effects, which may be relevant in the ground state, are, by necessity, neglected here. While their relevance is difficult to quantify, independent calculations on related Fe–Mn–Al–C κ carbides [46] suggest that the effect of magnetism on structural properties is limited, especially if qualitative trends are considered.

3. Results

3.1. Small Angle Neutron Scattering (SANS)

The model-based analysis of the ex-situ SANS scattering patterns using SASfit [47] reveals the presence of Mn–C SRO a few nanometers in size. The evolution of the mean cluster radius and number density of clusters per cubic centimeter was analyzed. The results showed that with the increase in recrystallization annealing time from 2 min to 30 min, the size of the Mn–C SRO decreased, while the number density increased. Figure 2 shows the SANS scattering curves of the X60Mn18 steel recrystallization annealed at 800 °C for 2 min and 30 min. The average cluster radius decreased from 2.10 nm (2 min) to 1.77 nm at the end of 30 min recrystallization annealing.

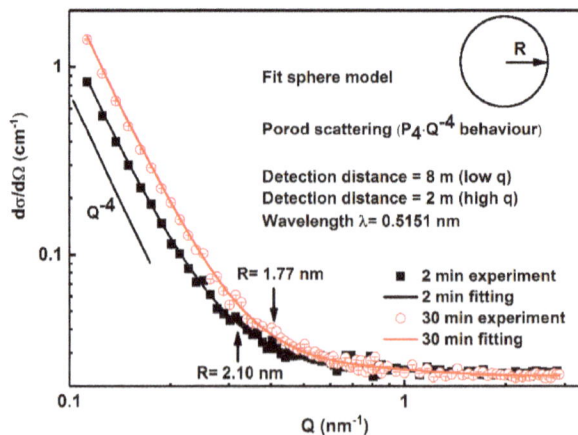

Figure 2. The small angle neutron scattering (SANS) curves of the X60Mn18 steel recrystallization annealed at 800 °C for 2 min and 30 min.

The evolution of the Mn–C SRO volume fraction distributions for the 2 min and 30 min annealed samples is shown in Figure 3. An increment in the annealing duration of non-deformed samples caused: (i) a shift of the cluster volume distribution curves to areas with smaller cluster sizes and (ii) a rise in the volume distribution peak height. With 30 min annealing, the clusters coarser than ≈2.15 nm disappeared, and the cluster mean radius, i.e., a cluster radius where the volume fraction reached its maximum, decreased from ≈2.10 nm (2 min) to ≈1.77 nm (30 min). The cluster volume fraction increased with the increase in recrystallization annealing time from 2 min to 30 min. The longer annealing time, i.e., 30 min annealing, brought an increment in the number density of clusters from 2.82×10^{14} cm^{-3} (2 min) to 7.43×10^{14} cm^{-3} (30 min).

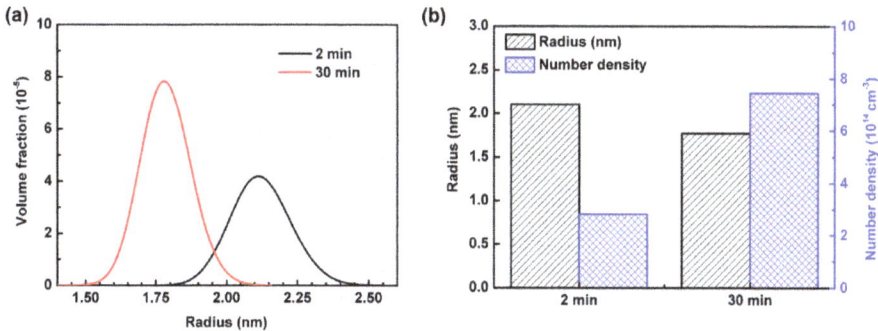

Figure 3. The evolution of Mn–C short-range ordering (SRO) volume fraction distribution of the X60Mn18 steel recrystallization annealed at 800 °C for 2 min and 30 min.

The annealing of an alloy system containing precipitations mostly resulted in a diffusion-controlled growth. However, in the present study, an increase in the duration during recrystallization annealing at 800 °C from 2 min to 30 min led to a decrease in mean particle size, and an increment in particle number density and volume fraction, which indicated new clustering formation and a finer distribution of the particles with smaller size. This new clustering response to the increasing annealing duration can be attributed to the recrystallization of the deformed microstructure, and enhanced diffusion. The deformed microstructure provided considerable defects in the material for the nucleation of the particles. The enhanced diffusion, i.e., longer diffusion duration, allowed atomic species to distribute within the solute solution approaching to equilibrium; in other words, it improved the elimination of segregations. Increasing recrystallization annealing duration gives rise to the observation of the microstructure closer to the equilibrium conditions, which was indicated by ab initio calculations in the present study.

3.2. Ab Initio Calculations

As a full sampling of all possible distributions of three C atoms among the 27 interstitial sites of the supercell is extremely demanding in terms of sheer computation, even when considering symmetry, a constraint for selection of the probable C positions had to be introduced. This was done under the justified assumption that the immediate chemical environment of the surrounding metal matrix, including the 1st and 2nd CS, is the driving criterion for the selection of a site, with C–C interactions among neighboring interstitials playing only a secondary role. While this is an approximation, it was considered to be reasonable based on prior results [29,30] and known length scales for atomistic SRO phenomena. Furthermore, carbon-vacancy interactions were not considered, as the modeled metal lattice is defect-free, thus containing no vacancies, whereas for the interstitial sites, the carbon/empty site fraction is $3/27 = 1/9$, so that empty voids are always present as 3rd CS neighbors for C in any case, rendering a distinction meaningless. Thus, in a first step, the 27 individual interstitial positions of each of the four structural models were occupied with only one C atom per cell to identify the energetically most favorable arrangement depending on local ordering in relation to the metal matrix, resulting in 4×27 distinct systems.

Figure 4 shows the total energies for each possible configuration averaged for configurations with similar ordering, $\overline{E}_{tot}$, in each STM as a function of the amount of Mn neighbors in the 1st CS, i.e., the surrounding octahedron, for the C atom, at a distance of roughly $\overline{r}_{C-Mn,1CS} \approx 1.90$ Å on average. This directly related to the choice of interstitial site as detailed above and was a sensible method of grouping the total energies, which are in general comparable for two systems with similar local ordering. However, for a random, disordered structure, a broader variation in energy could be expected due to the irregularity of the surrounding matrix, resulting in many different configurations if the occupation of the 2nd CS, with a distance of roughly $\overline{r}_{C-Mn,2CS} \approx 3.05$ Å on average, was also considered. While not shown in Figure 4a, which displays the averaged energies, a comparatively broad scattering of the individual energy values occurred, up to $\Delta E \approx 0.1$ eV for the same amount of Mn neighbors in both the 1st and 2nd CS. Because of that, identifying evident trends was difficult for the disordered case but, generally speaking, both the increasing amount of Mn neighbors in the 1st CS and, conversely, their decrease in number in the 2nd CS seem to be favorable energetically (a lower energy signifies a favored system per convention in chemistry). This agrees with results from previous studies on Mn in the 1st and 2nd CS of C in pure Fe matrices [30].

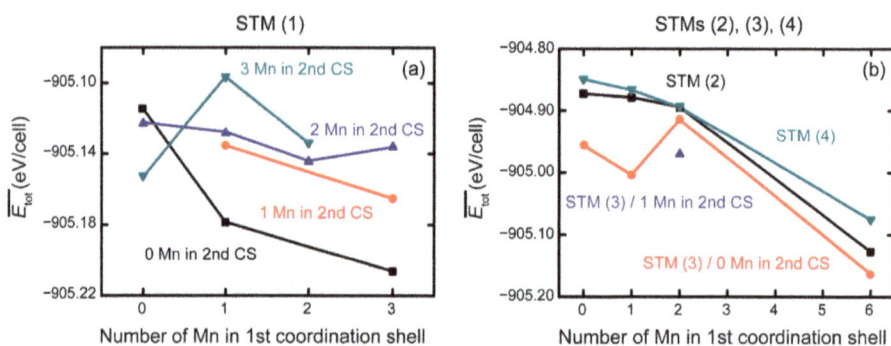

Figure 4. Averaged total energies $\overline{E}_{tot}$ as a function of the amount of Mn in the 1st coordination sphere (CS) of a single C atom in $Fe_{88}Mn_{20}C$ ($\overline{r}_{C-Mn,1CS} \approx 1.90$ Å). Lines between symbols are shown for better readability and are color-coded. (**a**) $\overline{E}_{tot}$ in the disordered structural model (STM) (1) with varying amounts of Mn atoms in the 2nd CS of C ($\overline{r}_{C-Mn,2CS} \approx 3.05$ Å); (**b**) $\overline{E}_{tot}$ in the ordered STMs (2), (3) and (4), with 0 and 1 Mn atoms in the 2nd CS for (3) and 0 for (2) and (4).

For the ordered STMs, these trends were more evident, as shown in Figure 4b. The decrease in $\overline{E}_{tot}$ with increasing Mn content in the 1st CS was valid for all three ordered variants. In contrast to the disordered STM, the individual energies (not shown) are much more narrowly distributed, in most cases with $\Delta E \approx 1$ meV, with the highest difference being $\Delta E \approx 0.04$ eV. An outlier was prominent: in STM (3), a system containing two direct Mn–C contacts was disfavored, contrary to the general trend (as the only system with two Mn atoms as 1st CS neighbors and zero Mn atoms in the 2nd CS for STM (3), it was represented individually in the figure). In said system, C occupies a bridging position between two full Mn_6 octahedra. Apart from this case, the entire dataset showed good agreement with the aforementioned results.

Due to the ordering constraint, there was limited data for higher Mn contents in the 2nd CS, with only eight systems, all in STM (3), out of the 81 distinct ordered cases (3×27, for all structures and individual C positions) containing one Mn in the 2nd CS. All of these eight systems contain two Mn atoms in the 1st CS, converging to a single data point in Figure 4b. Contrary to the disordered case, this led to a decrease in energy when compared to systems in STM (3) with zero Mn neighbors in the 2nd CS. However, due to the limited data mentioned above, this was not a statistically relevant observation and was likely caused by an interplay of other ordering effects for this particular arrangement.

In summary, the general trends (attractive Mn–C interactions for contacts in the 1st CS of C, repulsive interactions for those in the 2nd CS) were much less clearly present than in an ideally "diluted" system with an isolated Mn_6C octahedron in a Fe matrix, but could still be identified. The larger scattering of energy values for STM (1) suggested that longer-range interactions beyond the first two CS, which are especially present in the disordered STM due to a more homogenous distribution of Mn atoms throughout the host matrix, play a role as well. While these results generally agreed with earlier findings, the evident differences did stress both the importance of a large statistical sampling of different configurations when "realistic" models are employed (which was not fully realized in this study) and the limitations of the prior idealized model.

After this general analysis, the most energetically favorable arrangements were used to construct the $Fe_{88}Mn_{20}C_3$ structure approximating the real sample. While the total energy of the system was the primary selection criterion, the distance between C positions was also considered, resulting in the construction of several variants of $Fe_{88}Mn_{20}C_3$ systems for each of the four main STMs: four each for (1) and (2), and three each for (3) and (4). In order to investigate the magnitude of the repulsive C–C interaction, in several of these variants an energetically less favored position was nonetheless selected as a C site if the C–C distance increased as a result. The total energy of these systems was then evaluated.

Figure 5 shows the averaged total energies $\overline{E}_{tot}$ of the various $Fe_{88}Mn_{20}C_3$ systems as a function of the smallest C–C distance $r_{C–C,min}$ present in the respective structure. It is evident that a larger C–C distance is energetically favorable, which is shown clearly for STMs (1) and (2). The strong increase in energy for $r_{C–C,min} \approx 3.55$ Å in structure (2) in particular confirmed this general trend. For structures (3) and (4), the smallest $r_{C–C,min}$ is always approx. 3.77 Å, making a comparison difficult. However, the three variants for these structures each differ in C distribution, enabling an analysis of the C–C radial distribution functions (RDF) performed via the wxDragon package [48].

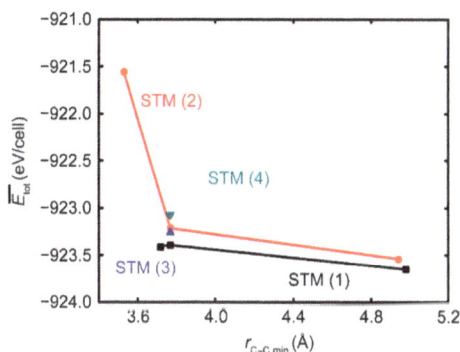

Figure 5. Averaged total energies $\overline{E}_{tot}$ as a function of the smallest C–C distance $r_{C–C,min}$ in the $Fe_{88}Mn_{20}C_3$ systems for all four structural models (STM). Lines between symbols are shown for better readability and are color-coded.

The respective RDF for the individual variant systems of structures (3) and (4) are shown in Figure 6. Two factors are evident: first, in the energetically favored structures, the number of Mn contacts in the 1st CS of C was higher than in the less favored ones, as expected due to Mn–C attraction. Second, the energetically favored arrangement in (3) exhibited a more homogenous distribution with C–C distances of approx. 3.77 Å, 4.89 Å, and 6.2 Å distributed roughly evenly (if differences in weight due to binning are considered), while the disfavored arrangement showed closer groupings of C atoms with $r_{C-C} \approx 3.77$ Å and 4.89 Å. However, as this was observed in one structure only, it conceivably might be a statistical artifact, and is in any case difficult to separate from the effect of the Mn–C interactions as outlined.

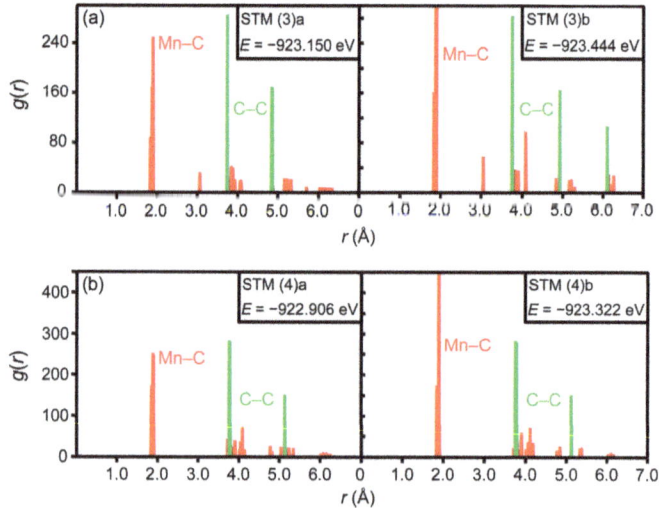

Figure 6. Radial distribution functions (RDF) for Mn–C (red) and C–C (green) contacts in $Fe_{88}Mn_{20}C_3$. (**a**) RDF for structural model (STM) (3), with a and b being the least and most energetically favored structures, respectively; (**b**) RDF for STM (4), likewise.

These combined findings conclusively demonstrated that a larger C–C distance is favorable and, in comparison, may compete with and even outweigh the effect of increasing the amount of Mn–C contacts in the 1st CS. This was reinforced by the fact that none of the energetically most favored arrangements for the $Fe_{88}Mn_{20}C_3$ variants corresponded to the selection of the respective three "best" single C positions obtained in the first step of the ab initio study. This suggests that, while Mn_6C octahedra were energetically preferred, they would not be located adjacent to each other, but rather distributed evenly. Thus, in real samples at the macroscopic scale, C atoms will preferably agglomerate in Mn-rich regions, but keep a minimal C–C distance of >5 Å at least, with an even spatial distribution. It should be noted that for more detailed quantification of these effects, follow-up studies with larger models and a broader statistical sampling are necessary, which have a low tractability for atomistic ab initio methods and are outside the scope of this mainly qualitative work.

3.3. Microstructure and Mechanical Properties

The microstructure of the X60Mn18 steel at different states is shown in Figure 7. The heavily deformed austenitic microstructure after cold rolling is shown in Figure 7a. The steel was fully recrystallized at 800 °C, both for 2 min and 30 min, as shown in Figure 7b,c. The austenitic grain size in the 2 min and 30 min samples was 8 µm and 15 µm, respectively. The DSA effect as manifested by the serrations in the stress-strain curves is shown in Figure 8a. These type-A serrations, which are characterized by a steep rise in stress alternating with plateau-like features, were observed in the stress-strain curves in the X60Mn18 steel recrystallization annealed for 2 min and 30 min in Figure 8. The type-A serrations occur as a result of the generation and propagation of a PLC band [16]. As an indicator of the DSA effect, jerky flow is found as a distinct feature of the tensile curves. The observation of each serration peak in the stress-strain curve results from a new-born PLC band. The X60Mn18 steel recrystallization annealed for 2 min exhibited higher yield strength (368.92 MPa) and ultimate tensile strength (1010.38 MPa) than those in the 30 min annealed sample. The elongation of the X60Mn18 steel annealed for 2 min and 30 min at 800 °C was 42.13% and 52.02%, respectively.

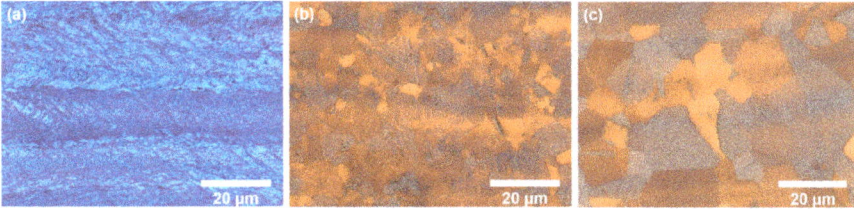

Figure 7. Microstructure of X60Mn18 steel (**a**) cold rolled; (**b**) recrystallization annealed at 800 °C for 2 min; (**c**) recrystallization annealed at 800 °C for 30 min.

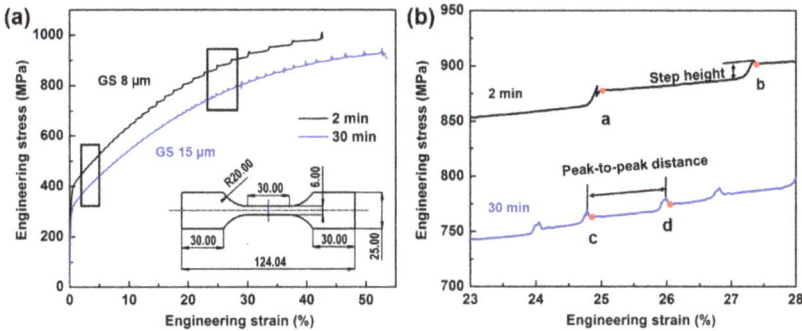

Figure 8. (**a**) Engineering stress-strain curves of 2 min (black) and 30 min (blue) annealed specimens; (**b**) The enlargement of the boxed segment of the strain-stress curve exhibiting type-A serrations, showing the serration with a steep rise in stress alternating by plateau-like features.

The partial magnified engineering stress-strain curves in Figure 8b revealed the serration behaviors of the steel. Compared with the 2 min annealed sample, the engineering stress-strain curve of the 30 min annealed sample comprised a higher number of serration peaks, i.e., a shorter peak-to-peak distance, with lower step heights. Within the deformation strain range from 23% to 28%, the peak-to-peak distance of the 2 min annealed sample was approximately 2.4% with a serration step height of 16.75 MPa, whereas the average peak-to-peak distance of the 30 min annealed sample was about 0.9% and the average serration step height was 8.70 MPa.

4. Discussion

Both the SANS measurements and the ab initio calculations reveal that, with an increase in the duration of recrystallization annealing of the X60Mn18 steel, the Mn–C units exhibiting SRO tend to keep distance towards each other and are more evenly distributed throughout the material, while exhibiting smaller size. The SANS results show that an increase in the duration of recrystallization annealing at 800 °C from 2 min to 30 min leads to a decrease in mean particle size and an increment of both particle number density and volume fraction, which indicates new clustering formation and a finer distribution of the particles with smaller size. The ab initio calculations show that thermodynamically, individual Mn_6C octahedra preferably keep minimum distances from each other (approximately 5 Å at least). The repulsive C–C interaction is an important effect, which can counteract the Mn–C attraction responsible for the formation of Mn_6C units in the first place, resulting in the above-mentioned units being evenly distributed rather than closely grouped. The DFT calculations thus explain the interactions for nearest-neighbor and second-nearest-neighbor Mn–C pairs and illustrate their SRO, while also clarifying the role of the C–C interaction on the atomic level. However, in reality, Mn and C do not form isolated octahedra only, but may also form clusters on a larger (nanometer) scale, as the SRO and SRC measured by SANS reveal. This suggests that the particles tend to show a dispersed, even distribution over long time annealing, explaining why, during recrystallization annealing, the Mn–C SRO tend to be more homogenously distributed with smaller particle size, higher number density and higher volume fraction. All in all, the SANS measurements and ab initio calculations of the Mn–C ordering, while examining different size scales, show good agreement with each other and suggest a possible underlying atomistic cause for cluster dispersion.

As shown in Figure 8, the serration phenomenon of investigated high-Mn steel underlines the DSA effect during plastic deformation. It was reported that the DSA in C-enriched high-Mn steels is attributed to the formation of Mn–C SRO [17,49,50]. Dastur and Leslie [17] claimed that the Mn–C octahedral clusters may reorient themselves in the stress field near a dislocation core, which strongly pins the dislocations and causes them to pile up. However, the pinning force is not strong enough to lock the mobile dislocations solidly and the pinned dislocations are released from the Mn–C SRO at once and propagate. This clustering leads to a higher lattice resistance to the dislocation glide, as the passage of a partial dislocation will in general change the local position of both substitutional and interstitial atoms [4]. The 2D composition-dependent stacking fault energy (SFE) map in Figure 9a summarizes the reported steels in the Fe–Mn–C system that exhibit serration flow during deformation. One can see that serration flow occurs in the high-Mn TRIP/TWIP steels of the Fe–Mn–C system that cover a wide range of chemical compositions [3,8,50–59].

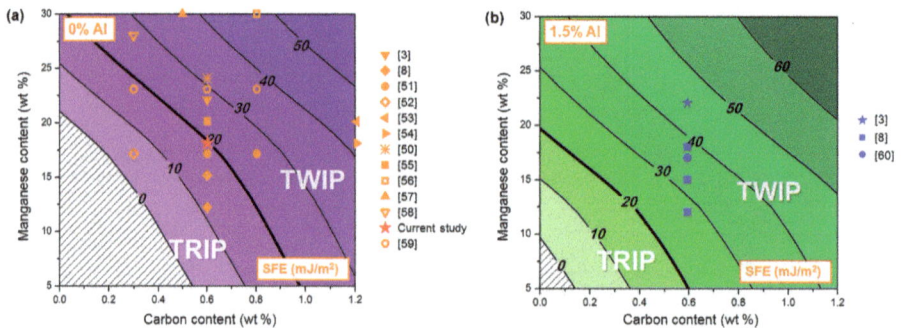

Figure 9. Two-dimensional (2D) composition-dependent stacking fault energy (SFE) maps summarizing the chemical composition of high-Mn steels with serration flow: (**a**) Fe–Mn–C system (**b**) Fe–Mn–1.5Al–C system.

The occurrence of serration requires the deformation level of the materials to meet a critical strain and the critical strain (ε_c) must first be achieved before serrated flow is observed. The critical strain for triggering serrations is influenced by SFE, deformation temperature, strain rate and elemental content (i.e., Al). The critical strain for the onset of the serrations shows an increasing trend with increasing SFE. An increase in the temperature promotes the serrations in the early stages of the deformation [16], the same as a lowered strain rate. The critical strain increases with increasing Al content. Al addition shifts the occurrence of DSA-related serrations towards higher strains or eliminates the serrations [11,60,61]. The steels in the Fe–Mn–1.5Al–C system that show serration flow during deformation are summarized in the 2D composition-dependent SFE map (Figure 9b). In the Fe–Mn–1.5Al–C steel system, the serrations reported in the literature occur mainly in the steels with 0.6 wt % carbon content. In the compositional range for high-Mn steels in the Fe–Mn–Al–C system, the serrations take place at higher deformation levels, which yields the required activation energy for C reorientation [13]. On the one hand, it was reported that Al reduces the activity and diffusivity of C [62,63] and delays the occurrence of serrations during deformation. On the other hand, ab initio calculations suggest that Al has a strong impact on the Mn–C SRO formation [30]. Al is preferable as a second-nearest neighbor to carbon; however, the attractive interaction of Mn–C pairs is affected by the presence of Al [30]. Because of high Al content in the system, the formation of Mn–C pairs is not preferred any longer [30]. In other words, the addition of a certain amount of Al suppresses the Mn–C SRO formation in high-Mn steels. This might be the reason for the delay or absence of the serration phenomenon in Al-alloyed high-Mn steels.

5. Conclusions

The ex-situ small angle neutron scattering (SANS) investigations prove the presence of Mn–C SRO in the recrystallization annealed X60Mn18 steel.

(1) With an increase in annealing time from 2 min to 30 min, the size of the Mn–C SRO decreases, whereas their number density increases. The material exhibits a more evenly dispersed distribution of smaller clusters at longer annealing time during recrystallization.

(2) The ab initio calculations qualitatively demonstrate that an increase in Mn neighbors of C in the 1st coordination sphere, resulting in a Mn_6C octahedron, is energetically favorable, confirming prior studies on the Mn–C interaction. However, in non-idealized, "realistic" models, these effects are far less pronounced and energetic scattering is higher.

(3) The repulsive C–C interaction is an equally important effect, which can counteract the Mn–C attraction, resulting in the formation of Mn_6C units that are distributed evenly rather than being closely grouped, and they keep minimum distances from each other. This confirms and expands upon the results of prior studies, which were restricted to idealized model systems and did not investigate the competition between the two effects.

(4) The ab initio calculations well describe the energetically favored condition of Mn–C clustering and provide a theoretical explanation on an atomistic scale of the clustering formation and evolution in the X60Mn18 steel.

Acknowledgments: This study was performed in the Collaborative Research Center (SFB) 761 "Steel ab initio" funded by the Deutsche Forschungsgemeinschaft (DFG), whose financial support is gratefully acknowledged. We thank the IT Centre of RWTH Aachen University and the Jülich-Aachen Research Alliance (JARA) HPC division for providing computational time and resources for the ab initio part of this work within the grant JARA0058. The experimental support from the KWS-2 beamline of Jülich Centre for Neutron Science (JCNS) at Heinz Maier-Leibnitz Zentrum (MLZ) is gratefully acknowledged.

Author Contributions: Wenwen Song designed the experiments and supervised the work. Wenwen Song, Ahmet Bahadir Yildiz and Judith E. Houston conducted the SANS experiments and analyzed the SANS data; Dimitri Bogdanovski conceived, performed and evaluated ab initio calculations; Richard Dronskowski and Wolfgang Bleck contributed with ideas and intensive discussions. All authors contributed to the interpretation of the results and the writing of the final version of the manuscript.

Conflicts of Interest: The authors declare no conflict of interest.

References

1. De Cooman, B.C.; Estrin, Y.; Kim, S.K. Twinning-induced plasticity (TWIP) steels. *Acta Mater.* **2018**, *142*, 283–362. [CrossRef]

2. Bouaziz, O.; Allain, S.; Scott, C.P.; Cugy, P.; Barbier, D. High manganese austenitic twinning induced plasticity steels: A review of the microstructure properties relationships. *Curr. Opin. Solid State Mater. Sci.* **2011**, *15*, 141–168. [CrossRef]

3. Yang, H.K.; Zhang, Z.J.; Dong, F.Y.; Duan, Q.Q.; Zhang, Z.F. Strain rate effects on tensile deformation behaviors for Fe–22Mn–0.6C–(1.5Al) twinning-induced plasticity steel. *Mater. Sci. Eng. A* **2014**, *607*, 551–558. [CrossRef]

4. Chen, L.; Kim, H.-S.; Kim, S.-K.; De Cooman, B.C. Localized deformation due to Portevin-LeChatelier effect in 18Mn–0.6C TWIP austenitic Steel. *ISIJ Int.* **2007**, *47*, 1804–1812. [CrossRef]

5.	Güvenç, O.; Roters, F.; Hickel, T.; Bambach, M. ICME for crashworthiness of TWIP Steels: From ab initio to the crash performance. *JOM* **2015**, *67*, 120–128. [CrossRef]

6.	Bouaziz, O.; Zurob, H.; Huang, M. Driving force and logic of development of advanced high strength steels for automotive applications. *Steel Res. Int.* **2013**, *84*, 937–947. [CrossRef]

7.	Bleck, W.; Guo, X.; Ma, Y. The TRIP effect and its application in cold formable sheet steels. *Steel Res. Int.* **2017**, *88*, 1700218. [CrossRef]

8.	Kim, J.-K.; De Cooman, B.C. Stacking fault energy and deformation mechanisms in Fe–*x*Mn–0.6C–*y*Al TWIP steel. *Mater. Sci. Eng. A* **2016**, *676*, 216–231. [CrossRef]

9.	Kang, M.; Shin, E.; Woo, W.; Lee, Y.-K. Small-angle neutron scattering analysis of Mn–C clusters in high-manganese 18Mn–0.6C steel. *Mater. Charact.* **2014**, *96*, 40–45. [CrossRef]

10.	Kim, J.; Estrin, Y.; De Cooman, B.C. Application of a dislocation density-based constitutive model to Al-alloyed TWIP steel. *Metall. Mater. Trans. A* **2013**, *44*, 4168–4182. [CrossRef]

11.	Jin, J.-E.; Lee, Y.-K. Effects of Al on microstructure and tensile properties of C-bearing high Mn TWIP steel. *Acta Mater.* **2012**, *60*, 1680–1688. [CrossRef]

12.	Jung, I.-C.; De Cooman, B.C. Temperature dependence of the flow stress of Fe–18Mn–0.6C–*x*Al twinning-induced plasticity steel. *Acta Mater.* **2013**, *61*, 6724–6735. [CrossRef]

13.	Lee, S.-J.; Kim, J.; Kane, S.N.; De Cooman, B.C. On the origin of dynamic strain aging in twinning-induced plasticity steels. *Acta Mater.* **2011**, *59*, 6809–6819. [CrossRef]

14.	Scott, C.; Allain, S.; Faral, M.; Guelton, N. The development of a new Fe–Mn–C austenitic steel for automotive applications. *Metall. Res. Technol.* **2006**, *103*, 293–302. [CrossRef]

15.	Gutierrez-Urrutia, I.; Raabe, D. Grain size effect on strain hardening in twinning-induced plasticity steels. *Scr. Mater.* **2012**, *66*, 992–996. [CrossRef]

16.	Kim, J.-K.; Chen, L.; Kim, H.-S.; Kim, S.-K.; Estrin, Y.; De Cooman, B.C. On the tensile behavior of high-manganese twinning-induced plasticity steel. *Metall. Mater. Trans. A* **2009**, *40*, 3147–3158. [CrossRef]

17.	Dastur, Y.N.; Leslie, W.C. Mechanism of work hardening in Hadfield manganese steel. *Metall. Trans. A* **1981**, *12*, 749–759. [CrossRef]

18.	Marceau, R.K.W.; Choi, P.; Raabe, D. Understanding the detection of carbon in austenitic high-Mn steel using atom probe tomography. *Ultramicroscopy* **2013**, *132*, 239–247. [CrossRef] [PubMed]

19.	Hellman, O.C.; du Rivage, J.B.; Seidman, D.N. Efficient sampling for three-dimensional atom probe microscopy data. *Ultramicroscopy* **2003**, *95*, 199–205. [CrossRef]

20.	Svergun, D.I.; Koch, M.H.J. Small-angle scattering studies of biological macromolecules in solution. *Rep. Prog. Phys.* **2003**, *66*, 1735–1782. [CrossRef]

21.	Radulescu, A.; Mathers, R.T.; Coates, G.W.; Richter, D.; Fetters, L.J. A SANS study of the self-assembly in solution of syndiotactic polypropylene homopolymers, syndiotactic polypropylene-block-poly(ethylene-co-propylene) diblock copolymers, and an alternating atactic-isotactic multisegment polypropylene. *Macromolecules* **2004**, *37*, 6962–6971. [CrossRef]

22.	Liu, Y.; Fratini, E.; Baglioni, P.; Chen, W.-R.; Chen, S.-H. Effective long-range attraction between protein molecules in solutions studied by small angle neutron scattering. *Phys. Rev. Lett.* **2005**, *95*, 118102. [CrossRef] [PubMed]

23.	Bambach, M.D.; Bleck, W.; Kramer, H.S.; Klein, M.; Eifler, D.; Beck, T.; Surm, H.; Zoch, H.-W.; Hoffmann, F.; Radulescu, A. Tailoring the hardening behavior of 18CrNiMo7-6 via Cu alloying. *Steel Res. Int.* **2016**, *87*, 550–561. [CrossRef]

24.	Song, W.; Radulescu, A.; Liu, L.; Bleck, W. Study on a high entropy alloy by high energy synchrotron X-Ray diffraction and small angle neutron scattering. *Steel Res. Int.* **2017**, *88*, 1700079. [CrossRef]

25.	Radlinski, A.; Mastalerz, M.; Hinde, A.; Hainbuchner, M.; Rauch, H.; Baron, M.; Lin, J.; Fan, L.; Thiyagarajan, P. Application of SAXS and SANS in evaluation of porosity, pore size distribution and surface area of coal. *Int. J. Coal Geol.* **2004**, *59*, 245–271. [CrossRef]

26.	Feigin, L.A.; Svergun, D.I.; Taylor, G.W. *Structure Analysis by Small-Angle X-Ray and Neutron Scattering*; Plenum Press: New York, NY, USA, 1987; pp. 83–87, ISBN 9781475766264.

27.	Radulescu, A.; Szekely, N.K.; Appavou, M.-S. KWS-2: Small angle scattering diffractometer. *J. Large-Scale Res. Facil.* **2015**, *1*, 29. [CrossRef]

28.	Pedersen, J.S. Determination of size distribution from small-angle scattering data for systems with effective hard-sphere interactions. *J. Appl. Crystallogr.* **1994**, *27*, 595–608. [CrossRef]

29. Von Appen, J.; Dronskowski, R. Carbon-induced ordering in manganese-rich austenite—A density-functional total-energy and chemical-bonding study. *Steel Res. Int.* **2011**, *82*, 101–107. [CrossRef]
30. Timmerscheidt, T.A.; Dronskowski, R. An ab initio study of carbon-induced ordering in austenitic Fe–Mn–Al–C alloys. *Steel Res. Int.* **2017**, *88*, 1600292. [CrossRef]
31. Bhadeshia, H.K.D.H. Carbon–carbon interactions in iron. *J. Mater. Sci.* **2004**, *39*, 3949–3955. [CrossRef]
32. Kresse, G.; Hafner, J. Ab initio molecular dynamics for liquid metals. *Phys. Rev. B* **1993**, *47*, 558–561. [CrossRef]
33. Kresse, G.; Furthmüller, J. Efficiency of ab-initio total energy calculations for metals and semiconductors using a plane-wave basis set. *Comput. Mater. Sci.* **1996**, *6*, 15–50. [CrossRef]
34. Blöchl, P.E. Projector augmented-wave method. *Phys. Rev. B* **1994**, *50*, 17953–17979. [CrossRef]
35. Kresse, G.; Joubert, D. From ultrasoft pseudopotentials to the projector augmented-wave method. *Phys. Rev. B* **1999**, *59*, 1758–1775. [CrossRef]
36. Perdew, J.P.; Burke, K.; Ernzerhof, M. Generalized gradient approximation made simple. *Phys. Rev. Lett.* **1996**, *77*, 3865–3868. [CrossRef] [PubMed]
37. Monkhorst, H.J.; Pack, J.D. Special points for Brillouin-zone integrations. *Phys. Rev. B* **1976**, *13*, 5188–5192. [CrossRef]
38. Methfessel, M.; Paxton, A.T. High-precision sampling for Brillouin-zone integration in metals. *Phys. Rev. B* **1989**, *40*, 3616–3621. [CrossRef]
39. Zunger, A.; Wie, S.-H.; Ferreira, L.G.; Bernard, J.E. Special quasirandom structures. *Phys. Rev. Lett.* **1990**, *65*, 353–356. [CrossRef] [PubMed]
40. Van de Walle, A.; Asta, M.; Ceder, G. The alloy theoretic automated toolkit: A user guide. *Calphad* **2002**, *26*, 539–553. [CrossRef]
41. Van de Walle, A.; Tiwary, P.; De Jong, M.; Olmsted, D.L.; Asta, M.; Dick, A.; Shin, D.; Wang, Y.; Chen, L.-Q.; Liu, Z.-K. Efficient stochastic generation of special quasirandom structures. *Calphad* **2013**, *42*, 13–18. [CrossRef]
42. Hafner, J. Ab-initio simulations of materials using VASP: Density-functional theory and beyond. *J. Comput. Chem.* **2008**, *29*, 2044–2078. [CrossRef] [PubMed]
43. Gyorffy, B.L.; Pindor, A.J.; Staunton, J.; Stocks, G.M.; Winter, H. A first-principles theory of ferromagnetic phase transitions in metals. *J. Phys. F Met. Phys.* **1985**, *15*, 1337–1386. [CrossRef]
44. Gonser, U.; Meechan, C.J.; Muir, A.H.; Wiedersich, H. Determination of Néel temperatures in fcc iron. *J. Appl. Phys.* **1963**, *34*, 2373–2378. [CrossRef]
45. Rankin, D.W. CRC handbook of chemistry and physics, 89th edition, edited by David R. Lide. *Crystallogr. Rev.* **2009**, *15*, 223–224. [CrossRef]
46. Dey, P.; Nazarov, R.; Dutta, B.; Yao, M.J.; Herbig, M.; Friák, M.; Hickel, T.; Raabe, D.; Neugebauer, J. Ab initio explanation of disorder and off-stoichiometry in Fe–Mn–Al–C κ carbides. *Phys. Rev. B* **2017**, *95*, 104108. [CrossRef]
47. Breßler, I.; Kohlbrecher, J.; Thünemann, A.F. SASfit: A tool for small-angle scattering data analysis using a library of analytical expressions. *J. Appl. Crystallogr.* **2015**, *48*, 1587–1598. [CrossRef] [PubMed]
48. Eck, B. *wxDragon Version 2.1.0*; RWTH Aachen: Aachen, Germany, 2016.
49. Owen, W.S.; Grujicic, M. Strain aging of austenitic hadfield manganese steel. *Acta Mater.* **1999**, *47*, 111–126. [CrossRef]
50. Saeed-Akbari, A.; Mosecker, L.; Schwedt, A.; Bleck, W. Characterization and prediction of flow behavior in high-manganese twinning induced plasticity steels: Part I. Mechanism Maps and Work-Hardening Behavior. *Metall. Mater. Trans. A* **2012**, *43*, 1688–1704. [CrossRef]
51. Koyama, M.; Sawaguchi, T.; Lee, T.; Lee, C.S.; Tsuzaki, K. Work hardening associated with ε-martensitic transformation, deformation twinning and dynamic strain aging in Fe–17Mn–0.6C and Fe–17Mn–0.8C TWIP steels. *Mater. Sci. Eng. A* **2011**, *528*, 7310–7316. [CrossRef]
52. Koyama, M.; Sawaguchi, T.; Tsuzaki, K. Influence of dislocation separation on dynamic strain aging in a Fe–Mn–C austenitic steel. *Mater. Trans.* **2012**, *53*, 546–552. [CrossRef]
53. Renard, K.; Ryelandt, S.; Jacques, P.J. Characterisation of the Portevin-Le Châtelier effect affecting an austenitic TWIP steel based on digital image correlation. *Mater. Sci. Eng. A* **2010**, *527*, 2969–2977. [CrossRef]
54. Koyama, M.; Sawaguchi, T.; Tsuzaki, K. TWIP effect and plastic instability condition in an Fe–Mn–C austenitic steel. *ISIJ Int.* **2013**, *53*, 323–329. [CrossRef]

55. Shen, Y.F.; Jia, N.; Misra, R.; Zuo, L. Softening behavior by excessive twinning and adiabatic heating at high strain rate in a Fe–20Mn–0.6C TWIP steel. *Acta Mater.* **2016**, *103*, 229–242. [CrossRef]
56. Lai, H.J.; Wan, C.M. The study of work hardening in Fe–Mn–Al–C alloys. *J. Mater. Sci.* **1989**, *24*, 2449–2453. [CrossRef]
57. Wang, X.; Zurob, H.S.; Embury, J.D.; Ren, X.; Yakubtsov, I. Microstructural features controlling the deformation and recrystallization behaviour Fe–30% Mn and Fe–30% Mn–0.5% C. *Mater. Sci. Eng. A* **2010**, *527*, 3785–3791. [CrossRef]
58. Song, W.; Ingendahl, T.; Bleck, W. Control of strain hardening behavior in high-Mn austenitic steels. *Acta Metall. Sin. (Engl. Lett.)* **2014**, *27*, 546–556. [CrossRef]
59. Song, W. Mn-C short-range ordering and local deformation in a high-manganese steels. *Mater. Sci. Eng. A* **2018**. in submission.
60. Daamen, M.; Nessen, W.; Pinard, P.T.; Richter, S.; Schwedt, A.; Hirt, G. Deformation behavior of high-manganese TWIP steels produced by twin-roll strip casting. *Procedia Eng.* **2014**, *81*, 1535–1540. [CrossRef]
61. Ma, Y.; Song, W.; Bleck, W. Investigation of the microstructure evolution in a Fe–17Mn–1.5Al–0.3C Steel via in situ synchrotron X-ray diffraction during a tensile test. *Materials* **2017**, *10*. [CrossRef]
62. Shun, T.; Wan, C.M.; Byrne, J.G. A study of work hardening in austenitic Fe–Mn–C and Fe–Mn–Al–C alloys. *Acta Metall. Mater.* **1992**, *40*, 3407–3412. [CrossRef]
63. Zuidema, B.K.; Subramanyam, D.K.; Leslie, W.C. The effect of aluminum on the work hardening and wear resistance of hadfield manganese steel. *Metall. Trans. A* **1987**, *18*, 1629–1639. [CrossRef]

![metals logo] *metals*

MDPI

Article

Linking Ab Initio Data on Hydrogen and Carbon in Steel to Statistical and Continuum Descriptions

Marc Weikamp [1], Claas Hüter [1,2,*] and Robert Spatschek [1,2]

[1] Forschungszentrum Jülich GmbH, Institute for Energy and Climate Research, IEK-2, 52428 Jülich, Germany; m.weikamp@fz-juelich.de (M.W.); r.spatschek@fz-juelich.de (R.S.)

[2] Jülich-Aachen Research Alliance (JARA Energy), RWTH Aachen University, 52056 Aachen, Germany

* Correspondence: c.hueter@fz-juelich.de; Tel.: +49-2461-61-1569

Received: 28 February 2018; Accepted: 23 March 2018; Published: 27 March 2018

Abstract: We present a selection of scale transfer approaches from the electronic to the continuum regime for topics relevant to hydrogen embrittlement. With a focus on grain boundary related hydrogen embrittlement, we discuss the scale transfer for the dependence of the carbon solution behavior in steel on elastic effects and the hydrogen solution in austenitic bulk regions depending on Al content. We introduce an approximative scheme to estimate grain boundary energies for varying carbon and hydrogen population. We employ this approach for a discussion of the suppressing influence of Al on the substitution of carbon with hydrogen at grain boundaries, which is an assumed mechanism for grain boundary hydrogen embrittlement. Finally, we discuss the dependence of hydride formation on the grain boundary stiffness.

Keywords: hydrogen embrittlement; ab initio; scale bridging

---

1. Introduction

One of the challenges for the modeling of materials, like metals and steel, is that a description on a single length scale is not sufficient in many cases. The bridging of scales, however, is not a routine task and requires to accurately link expressions of the different modeling levels. Nowadays, ab initio methods are able to deliver a large variety of material parameters and process routes with high accuracy, but are confined to systems with about ~100 atoms at most. At the other extreme, continuum methods are able to capture e.g., long ranged elastic effects, but they require material parameters from the lower scales. In between, statistical and thermodynamic perspectives can help to bridge the scales, and to link also equilibrium to out-of-equilibrium regimes. Often, the parameters to be transferred across the scales involve some aspects of coarse graining, as the level of detailed (atomistic) information is lost on the higher scales.

Here, we demonstrate how selected scale transitions between the atomic and continuum scale can be performed, also in a quantitative manner. All these aspects are related to failure mechanisms in high and medium manganese steel, with a special focus on hydrogen. Especially during the last 20 years, while isolated scale bridging investigations on this topic have been conducted, the general need for a comprising scale bridging approach became increasingly clear when light could be shed onto the complex interplay of effects on multiple length and time scales [1–3]. Nevertheless, we keep some aspects more generic and discuss e.g., also carbon as interstitial element. The following four features are discussed within this article:

- We demonstrate that volume dependent solution enthalpies of carbon in iron are fully consistent with the theory of linear elasticity.
- We show how solution enthalpies of hydrogen can be used for statistical considerations, and how aluminum affects the occupation of octahedral sites in austenitic steel.

- We introduce a simple mean field scheme to estimate energies of a grain boundary populated with hydrogen and carbon and use that to discuss hydrogen enrichment at carbon saturated grain boundaries in ferrite and the role of aluminum as protective agent.
- We propose a simple model for grain boundary stiffening or weakening, which affects the probability for hydride formation near grain boundaries in comparison to the bulk.

These applications cover different aspects of importance to hydrogen embrittlement and illustrate how to establish a quantitative link between ab initio simulations, statistical descriptions and continuum theories.

2. Scale Bridging for Strain Dependent Solution Enthalpies in Fe-C

In this section, we demonstrate how data, which is obtained from ab initio or empirical potential calculations, can be transferred to the continuum level. Usually, this transfer is not straightforward, and we use here a particularly simple case to demonstrate this.

The starting point is published data on the solution enthalpy of carbon in bcc iron [4]. This paper contains a rich collection of strain dependent energies for different interatomic potential descriptions, and it is useful to compare the predictions to a continuum perspective, which allows using the computed data also for larger scale modeling.

First, we note that C in bcc iron occupies octahedral positions in equilibrium, and induces a tetragonal distortion. This effect is considered in the careful ab initio based computations of Hristova et al. [4]. They additionally considered the situation where they neglect this anisotropy and assumed a cubic symmetry. It turns out that the results do not change significantly, if the full relaxation is considered. For the present purpose, it has the advantage that we can use the simpler cubic symmetry for the comparison, which leads to more compact analytical expressions on the continuum scale. In the end, we briefly discuss the transfer to the more general situation. In addition, we note that hydrogen in austenite, which is discussed in the other parts of this manuscript from the perspective of a scale bridging, does not exhibit such tetragonal distortions, and therefore the expressions can directly be transferred to this case.

For the calculation of hydrogen solution enthalpies, the authors of [4] use a $3 \times 3 \times 3$ supercell consisting of 54 iron atoms. For the numerical parameters related to the ab initio calculations, we refer to the description in [4]. One of the octahedral positions is occupied by a carbon atom. From the arising energy differences, the solution enthalpy is defined as

$$H^{exc}(V) = E^{Fe_{54}C}(V) - E^{Fe_{54}}(V) - E^C. \tag{1}$$

Here, all energies are $T = 0\,\mathrm{K}$ bare energies from the underlying simulations. $E^{Fe_{54}C}(V)$ is the volume dependent energy of the system with an additional carbon atom and $E^{Fe_{54}}(V)$ the same without the carbon. Notice that both expressions are taken at the same volume. We note that carbon wants to expand the lattice, therefore the first situation has a higher compressive strain as the latter. For a handshaking to continuum methods, it is important to take into account such effects, as we will see below. Finally, E^C is a reference potential from a "carbon reservoir". For the data related to density functional theory calculations (DFT), Hristova et al. use diamond as carbon reservoir [4]. As we are mainly interested here in the elastic rather than the chemical contributions, this reference term turns out to be of minor importance, but in general it is important to consider it e.g., for thermodynamic considerations.

On the continuum level, a natural approach for the total energy of the system is an additive decomposition into chemical and elastic contributions,

$$E(\epsilon_{ij}, c) = E_{el}(\epsilon_{ij}, c) + E_c(c), \tag{2}$$

where c is the carbon concentration, and ϵ_{ij} the strain tensor components, as defined below. A central difference between the discrete and continuum energies is that for large scale applications it is often more convenient to normalize the energy per volume rather than per particle. The reason is that, in more complex situations, where inhomogeneous strain and concentration distributions occur in non-equilibrium states, e.g., phase field approaches can be a versatile approach for modeling the time dependent microstructural evolution. For that, a generating energy functional is written as space-integral over an energy density, which is therefore defined with dimension energy per volume.

For the isotropic expansion (under the constraints mentioned above), the elastic energy can be expressed as

$$E_{el}(\epsilon_{ij}, c) = \frac{9}{2}BV_0(\epsilon_{ij} - \epsilon_{ij}^0(c))^2, \tag{3}$$

using the linear theory of elasticity and an isotropic volume expansion, as explained in detail below. Here, V_0 is the undeformed volume in the Lagrangian reference state, B the bulk modulus, and ϵ_{ij} are the linear elastic strain tensor components

$$\epsilon_{ij} = \frac{1}{2}\left(\frac{\partial u_i}{\partial x_j} + \frac{\partial u_j}{\partial x_i}\right), \tag{4}$$

expressed through the displacement field components u_j. Here, the strain is diagonal with $\epsilon_{ij} = \epsilon\delta_{ij}$ and $\delta_{ij} = 1$ for $i = j$ and 0 otherwise. The eigenstrain $\epsilon_{ij}^0(c)$ expresses the stress-free expansion of the lattice through the presence of a finite carbon concentration c. As long as the concentration is not too high, it is appropriate to assume Vegard's law, hence the eigenstrain scales linearly with c. In addition, if we neglect the tetragonal distortion, the expansion is uniform, hence the eigenstrain takes the tensorial form

$$\epsilon_{ij}^0 = \chi c\delta_{ij} = \epsilon_0\delta_{ij}, \tag{5}$$

with the Vegard coefficient χ. For the definition of the concentration, we can use c being the ratio of populated and total number of octahedral positions, hence c is in the range from 0 to 1.

In general, the bulk modulus is also concentration dependent. However, for low concentrations, this effect can typically be neglected. This is supported by the DFT data in [4]. Both with and without one carbon atom the bulk modulus is reported to be $B = 189$ GPa. Conceptually, a concentration dependence of the elastic constants leads to a higher order correction of the leading terms.

According to [4], the single carbon atom in the 54 atom host matrix leads to a volume increase of 1.7%, hence the strain in each direction is $1/3$ of this value, $\epsilon_0 = 0.0047$. The volume of the reference cell is 11.42 Angstrom^3 per iron atom.

The corresponding expression to the solution enthalpy (1) is in continuum representation

$$H^{exc} = E(\epsilon, c) - E(\epsilon, 0) - E^C \approx \left.\frac{\partial E_{el}(\epsilon, c)}{\partial c}\right|_{c=0} c + E_c'(0)c - E^C \tag{6}$$

after Taylor expansion. This gives

$$H^{exc} = -9BV_0\epsilon\chi c + E_c'(0)c - E^C = -9BV_0\epsilon\epsilon_0 + E_c'(0)c - E^C. \tag{7}$$

Under isotropic volume change the strain becomes $\epsilon = \frac{1}{3}(V/V_0 - 1)$, and, according to the above expression, the solution enthalpy should be a linear function of the strain or volume and decay with increasing volume. Interestingly, this behavior is qualitatively satisfied by the DFT data in [4], but not at all by several of the used empirical potentials (see Figure 1). The authors argue that these potentials are therefore unsuitable for describing the strain effects of hydrogen in iron. Here, we confirm this statement from a continuum perspective, as apparently these potentials disagree with predictions from elasticity theory. We can furthermore perform a quantitative comparison using the slope $H^{exc'}(V)$, which in the continuum picture becomes

$$\frac{dE^{exc}}{dV} = -3B\epsilon_0. \tag{8}$$

Inserting the material parameters given above gives the prediction $H^{exc'}(V) = -0.0201\,\text{eV}/\text{Angstrom}^3$, which is in perfect agreement with the DFT value (identical to the given accuracy, see Table 3 in [4]), and deviates from the empirical potentials, as expected. The continuum result for the strain dependence of the solution enthalpy is depicted in Figure 1. Only for large strains, where linear theory of elasticity is no longer applicable [5], deviations occur with the DFT results.

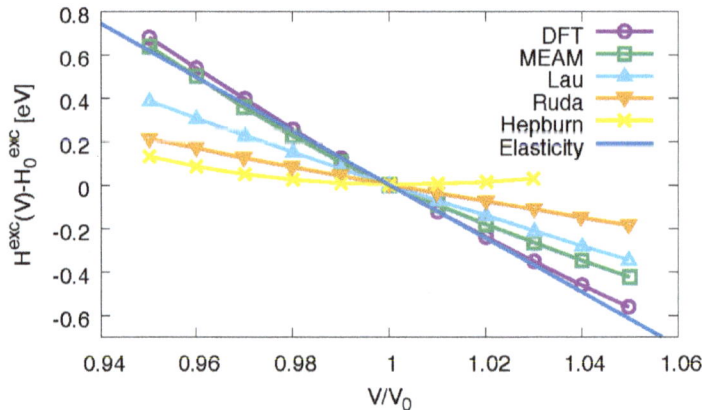

Figure 1. Solution (excess) enthalpy of carbon in bcc iron as function of volume. The continuum theory prediction is compared to DFT and empirical potential data. See [4] for details of the atomic scale simulations and the different empirical potentials.

Finally, we mention that consideration of tetragonality is a rather straightforward extension of the preceding discussion. For that, one uses the more general expression

$$E_{el} = \frac{1}{2}VC_{ijkl}(\epsilon_{ij} - \epsilon_{ij}^0)(\epsilon_{kl} - \epsilon_{kl}^0) \tag{9}$$

for the elastic energy, and the eigenstrain depends then on the carbon concentrations in the different sublattices. In practise, ϵ_{ij}^0 becomes a superposition of the tetragonal expansions in the different directions.

Overall, this example shows how the transfer from discrete atomistic descriptions to continuum pictures involving elastic effects can be done. From this, one can obtain a closed analytical expression with all material parameters, which can subsequently be used for higher scale modeling.

3. Hydrogen in Fe-Mn-Al Alloys

In this section, we discuss certain thermodynamic aspects of hydrogen in austenitic Fe-Mn-Al alloys. The starting point is the work by von Appen et al. on hydrogen solubility in Fe-Mn alloys [6]. Figure 2 shows a collection of the solution enthalpy of hydrogen. Similarly to the discussion above, it is defined as

$$H = E^{Fe_{1-x}Mn_xH} - E^{Fe_{1-x}Mn_x} - \mu_H, \tag{10}$$

where the reference chemical potential μ_H is half of the energy of an isolated H_2 molecule. Notice that the calculations were obtained for fully relaxed configurations, hence in particular the pressure vanishes, and consequently the volumes of the two energy terms with and without hydrogen differ. However, from the considerations above, we can conclude that the Legendre transformation between fixed volume and fixed pressure can be performed using theory of elasticity, without computationally expensive additional ab initio simulations. Obviously, the energy essentially depends linearly on the number of manganese neighbors around the H occupied octahedral site (see Figure 2). We note that the same linear energy trend is even more pronounced if the solution enthalpy is plotted versus the Voronoi volume. We can therefore conclude that the leading contribution is due to elastic effects.

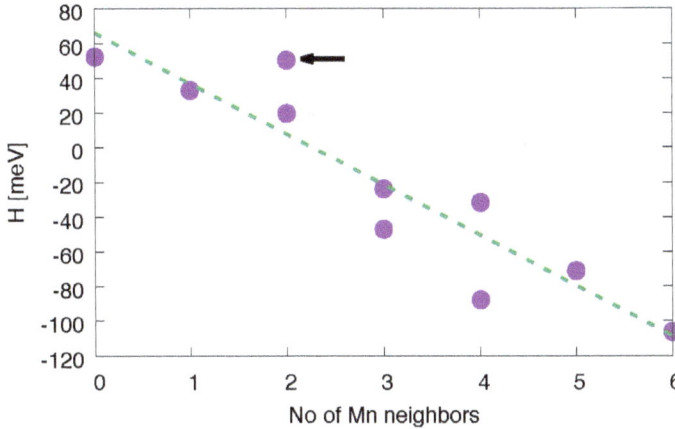

Figure 2. Hydrogen solution enthalpy as function of the number of Mn neighbors. For some cases, there are different energy values, which are due to different atomic arrangements, e.g., trans or cis configurations (see [6] for a detailed discussion). The dashed curve is a linear fit. The black arrow points to the trans configuration with two neighboring Mn atoms at opposite ends of the octahedron surrounding the H atom. This configuration is the reference for the Fe-Mn-Al-H system.

We can use this information to obtain some insights into the equilibrium occupation of the different octahedral sites by hydrogen. For an alloy with the composition $Fe_{1-x}Mn_x$, we assume a random distribution of Fe and Mn atoms, which ignores potential near ordering effects. Therefore, the probability for an octahedral site to have exactly n manganese neighbors is given by

$$p(n) = \binom{6}{n}(1-x)^{6-n}x^n. \tag{11}$$

The occupation of a single octahedral site follows a Fermi-Dirac distribution,

$$c(E) = \frac{1}{1 + \exp(\beta[E(n) - \mu])}, \tag{12}$$

where $0 < c(E) < 1$ is the probability of the site to be populated with hydrogen, $\beta = 1/kT$ is the inverse temperature, μ the equilibrium chemical potential of hydrogen (not to be confused with μ_H above), and E is the solution enthalpy for the given atomic environment. Therefore, the equilibrium hydrogen concentration in the Fe-Mn system becomes

$$c = \sum_{n=0}^{6}\binom{6}{n}(1-x)^{6-n}x^n\frac{1}{1 + \exp(\beta[E(n) - \mu])}. \tag{13}$$

As the concentrations are typically low, the additional consideration of more than one hydrogen atom in the supercell should not be necessary. With the linear approximation $E(n) = a - bn$ (see Figure 2, using $a = 66$ meV and $b = 29$ meV), we can relate the average concentration c to the chemical potential. If we consider (here, for example, for $x = 0.2$) $c = 0.5$ ppm at room temperature, which is a concentration that may already trigger hydrogen embrittlement, the equilibrium chemical potential is around $\mu = -351$ meV. An octahedron with 8 Mn neighbors is rare within such an alloy, but energetically most favorable. Its average population with hydrogen is around $c(n = 6) \approx 60$ ppm, which is two orders of magnitude higher than the average concentration c. Similarly, the energetically most unfavorable configuration, where hydrogen is surrounded by Fe only, has an expected H occupation of $c(n = 0) \approx 0.06$ ppm, which is one order of magnitude lower than the average value. The octahedral positions with a large number of Mn atoms may therefore be considered as hydrogen traps, in particular as their energy is about 130 meV lower than the average site. However, in equilibrium, even these sites have only a minute H occupation, and therefore these sites cannot trap a large amount of hydrogen. This is a general feature of hydrogen with such low concentrations. The chemical potential is very low, and hence all site occupations are in the tail $E - \mu \ll -kT$ of the Fermi-Dirac distribution. Typical trap depths are not sufficient to provide a significant increase of the hydrogen concentration beyond a dilute limit, which would require reaching the regime $E - \mu \sim -kT$.

Aluminum changes the energy landscape rather strongly, as worked out for specific atomic arrangements in [7,8]. Both author groups use different concentrations of their alloys, which is captured by a 32 atom fcc supercell. Song et al. use $Fe_{26}Mn_5Al_1$, Hüter et al. $Fe_{23}Mn_8Al_1$, and both groups come to similar conclusions despite the different stoichiometry, which suggests that the situation is more general. We follow the notation in [8], where *x-y-z* denotes an octahedral environment with *x* iron, *y* manganese and *z* aluminum atoms; this notation, however, cannot distinguish between different atomic arrangements, but we stay with it for convenience. In the following discussion, we focus on the $Fe_{23}Mn_8Al$ supercell. This system allows for distinguishing 6-0-0, 5-0-1, 4-2-0 and 3-2-1 neighbourhood configurations for the hydrogen atom.

The starting point is the trans (high energy) 4-2-0 configuration in Figure 2 (notice that the simple notation 4-2-0 does not reflect whether Al is present in a next-to-nearest neighbor position, therefore this information needs to be provided additionally). In contrast to the Fe-Mn alloys discussed above, the energy is here not only determined by the nearest neighbor configuration. A 4-2-0 neighborhood, where an Al atom occupies a next-to-nearest neighbor position, diminishes the hydrogen solution enthalpy by about 60 meV [7]. It is therefore speculated that this octahedral configuration may act as an H trap; in view of our discussion above, we stress that this effect is however rather weak.

More pronounced, however, is the effect that a direct neighborhood of Al to an H atom is energetically highly unfavorable. Configurations like 3-2-1 or 5-0-1 are about 130–140 meV less favorable than the potential 4-2-0 "trap" site. These octahedra are populated with a concentration that is about two orders of magnitude lower than the average concentration at room temperature; we may therefore effectively say that such sites are blocked.

Compared to the Fe-Mn system, Al therefore leads to strong alternations of the entire energy landscape. Changes of atomic neighborhoods by replacing one Fe atom by Al raises the energy in several cases by around 80 meV. In contrast, only one extreme case in the Fe-Mn system, for a transition between the trans 4-2-0 to the mere 3-3-0 environment, leads to a comparable shift of about -80 meV. In all other cases, the energetic shifts are much milder in the Fe-Mn system. Altogether, we can therefore conclude that the entire solution enthalpy landscape becomes significantly rougher in the Fe-Mn-Al case by the addition of rather small amounts of Al. Already for about 3 at% of Al almost 20% of all octahedral cages are neighboring an aluminum atom and are therefore energetically unfavorable for hydrogen.

4. Ferritic Grain Boundary Hydrogen Embrittlement: Estimates From Ab Initio Datasets

In this section, we discuss a possible phase separation into hydrogen enriched regions and hydrogen depleted regions at ferritic grain boundaries, and the function of aluminum as protective agent against grain boundary related hydrogen embrittlement.

An important contribution to the macroscopic interpretation of hydrogen embrittlement is the phase formation based on available ab initio data. Especially in the case of hydrogen in steel, this approach becomes complex, as at least carbon, hydrogen and iron need to be considered, and the required amount of ab initio data quickly becomes large. Especially for grain boundaries, which require rather expensive electronic scale simulations, the data set is sparse. Therefore, the development of possible approximative schemes can be helpful in this context.

Our first approach employs a mean field approximation for grain boundary energies, based on data published in [4,9,10] for varying carbon and hydrogen populations. The most investigated grain boundary in these publications is $\Sigma5(310)[001]$ in bcc iron, and the authors present—amongst other results—segregation and grain boundary energies for carbon and hydrogen at grain boundaries, where the most attractive type of grain boundary site is populated. In the considered supercell, four of these sites exist per grain boundary, i.e., any combination of altogether four hydrogen or carbon atoms per grain boundary corresponds to a fully saturated grain boundary with respect to this site. For details about the determination of the most attractive sites in a grain boundary state of minimal energy, we refer to [4,9–11]. The grain boundary energies reported in [9] are defined as

$$\gamma_{GB}^{nX} = \frac{E_{GB}^{Fe+2nX} - E_{bulk}^{Fe} - 2n\mu_X}{2A}, \tag{14}$$

where γ_{GB}^{nX} is the energy of a grain boundary with interfacial area A and n atoms of type X in it. E_{GB}^{Fe+2nX} is the absolute energy of the used simulation cell, which includes two grain boundaries of the simulated type due to periodicity constraints. By E_{bulk}^{Fe}, the absolute energy of a simulation cell with the same number of iron atoms is defined. Note that all simulations were relaxed using zero stress boundary conditions for the supercell, i.e., the only elastic contributions of the resulting energies are those confined to atomistic length scales around the grain boundary. μ_X is the reference chemical potential, for hydrogen that corresponds to H_2 molecules, for carbon to diamond, as discussed also in the preceding sections.

The scheme to estimate grain boundary energies for various hydrogen and carbon populations of the reported $\Sigma5(310)[001]$ grain boundaries is a summation of the averaged energetic contributions per atom relative to the fully saturated grain boundary for each pure population, which can be understood as a mean field approach. This can be expressed as

$$\gamma_{GB}^{nH+mC} = \frac{n}{4}\gamma_{GB}^{4H} + \frac{m}{4}\gamma_{GB}^{4C} + \frac{4-n-m}{4}\gamma_{GB}, \tag{15}$$

where the required grain boundary energies reported in [9] are $\gamma_{GB} = 1.55\,\text{Jm}^{-2}$, $\gamma_{GB}^{4H} = 1.27\,\text{Jm}^{-2}$, and $\gamma_{GB}^{4C} = 0.58\,\text{Jm}^{-2}$. Besides the reported grain boundary energies for zero to four sites occupied by carbon or hydrogen, also a co-segregated grain boundary energy is provided, for two hydrogen atoms and one carbon atom. This co-segregated grain boundary energy is an interesting test data point for the approximative scheme, which would predict the grain boundary energy for the reported population as 1/4 of a carbon saturated grain boundary, 2/4 of a hydrogen saturated grain boundary and 1/4 of an empty grain boundary. In Table 1 below, we see the comparison of predicted mean field grain boundary energies and ab initio calculated grain boundary energies for the testable data points in [9].

Apparently, the mean field description works excellently for hydrogen, while the discrepancies for carbon can reach up to 0.09 J/m². This has to be put in context of the variance in reported ab initio results for grain boundary energies that correspond to identical definitions and therefore should give identical numbers. We provide here the grain boundary energies reported for two grain boundaries. The reporting references include [9–15], and the reported grain boundary energies range from 1.49 Jm^{-2} to 1.63 Jm^{-2} ($\Sigma5(310)$) and from 0.43 Jm^{-2} to 0.47 Jm^{-2} ($\Sigma3(11\bar{2})$).

While the approximative scheme can be tested for the $\Sigma5(310)[001]$ grain boundary in the especially interesting regime of co-segregated grain boundaries, we are currently not aware of grain boundary energies for co-segregated sites for any other grain boundary. Consequently, we continue with a focus on the $\Sigma5(310)[001]$ grain boundary.

Table 1. Comparison of calculated and estimated grain boundary energies.

Grain Boundary	Calculated Value/(Jm^{-2})	Estimated Value/(Jm^{-2})	Error/(Jm^{-2})
γ_{GB}^{1H}	1.49	1.48	0.01
γ_{GB}^{2H}	1.42	1.41	0.01
γ_{GB}^{3H}	1.34	1.34	<0.01
γ_{GB}^{1C}	1.25	1.31	0.06
γ_{GB}^{2C}	0.90	0.93	0.03
γ_{GB}^{3C}	0.73	0.82	0.09
γ_{GB}^{2H1C}	1.08	1.17	0.09

The calculated grain boundary energies reported in [9] and their estimated values based on the mean field approximation. We do not compare the energies for four carbon and four hydrogen atoms as they coincide by definition of the approximation.

As the next step, we obtain the energies required for the determination of grain boundary phase separation. From the findings reported in [9], we know that the ground state of the grain boundary is carbon saturated with respect to the most attractive sites. These sites are also most attractive to hydrogen, thus the phase equilibrium in the ternary Fe-C-H system at the grain boundary reduces to the coexistence of regions in the grain boundary with varying hydrogen and carbon fractions, where all attractive sites are occupied by one of those species.

When the mixture of carbon saturated (hydrogen free) phase (α) and hydrogen saturated phase (β) forms, the averaged Gibbs energy per host atom is given as

$$g(x_H, T) = (1 - v)g_\alpha(x_\alpha, T) + vg_\beta(x_\beta, T), \tag{16}$$

where g_α is the averaged Gibbs energy per host atom in the hydrogen-free phase, g_β is the averaged Gibbs energy per host atom in the hydride phase, x_H, x_α, x_β are the averaged hydrogen concentration, hydrogen concentration in the carbon-saturated phase and hydrogen concentration in the hydride phase. We assume that we have an ideal statistical distribution of the hydrogen atoms along the available sites, and restrict the description to the configurational entropy contribution. The Gibbs energies of pure iron (α), $x_\alpha = 0$ and a pure hydride phase (β, fully saturated, $x_\beta = 1$) are

$$g_{\alpha,\beta} = g_{\alpha,\beta}^0(x_{\alpha,\beta}) + g_c(x_{\alpha,\beta}, T), \tag{17}$$

where, in the low concentration regime,

$$g_c \sim kT x_H \ln(x_H/x_0) \tag{18}$$

is the dominant contribution. We require equal chemical potentials for the heterogenous mixture of hydrogen saturated and carbon saturated grain boundary phases and the α-phase, so we obtain for low temperatures

$$\frac{\left[-g_\alpha(0) + g_\beta(x_\beta^0)x\right]}{x_\beta} \approx \frac{\partial g_\alpha(0)}{\partial x_\alpha^0} + kT \ln(x_H/x_0), \tag{19}$$

where we used $v \approx x_H/x_\beta^0$, with $x_{\alpha,\beta}^0$ denoting the respective phase equilibrium concentration. For the solubility limit, we get

$$x_H \approx x_0 \exp\left(-\frac{\Delta G}{kT}\right), \tag{20}$$

with $\Delta G = \left[-g_\alpha(0) + g_\beta(x_\beta) \right] / x_\beta^0 + \partial g_\alpha / \partial x_{x=0}$. The derivative of the Gibbs energy at zero concentration of hydrogen, $\partial g_\alpha / \partial x_{x=0}$, is obtained as finite difference approximation from the available ab initio data. The expression for the enthalpy difference ΔG becomes more familiar when we rewrite it in terms of classical defect energies $E^f_{\alpha,\beta}$, and we identify the relevant contributions for ΔG,

$$\Delta G = E^f_\alpha - E^f_\beta, \tag{21}$$
$$E^f_\alpha = E^{tot;3C:1H}_{GB;N_M} - E^{tot;4C}_{GB;N_M} - \mu^{ref}_H + \mu^{ref}_C,$$
$$E^f_\beta = E^{tot;1H}_{GB;1/x^0\beta} - E^{tot;1C}_{GB} - 1\mu^{ref}_H + 1\mu^{ref}_C,$$
$$E^f_\beta = \frac{1}{4}\left[E^{tot;4H}_{GB;N_M} - E^{tot;4C}_{GB} - 4\mu^{ref}_H + 4\mu^{ref}_C \right].$$

In the last line, we took advantage of the scaling behaviour of the energy with respect to the particle numbers. By $E^{tot;nC:mH}_{GB;N_M}$, we define the total energy of the grain boundary simulation cell including n carbon atoms and m hydrogen atoms at the attractive sites at each of the two grain boundaries, and N_M host atoms. To calculate the differences in energies per atom from the differences in grain boundary energies, $\Delta E = 3.207\,\text{eV}\,\text{m}^2\text{J}^{-1}\Delta\gamma_{GB} + \Delta\mu_{ref}$ holds for the given grain boundary area of $A = 51.371 \times 10^{-20}\,\text{m}^2$. The values obtained here for a separation in the considered $\Sigma 5(310)[001]$ grain boundary into fully carbon saturated and fully hydrogen saturated phases lead to a hydrogen solubility at the interface governed by the enthalpy difference ($\Delta G \approx 0.06\,\text{eV}$), where the reference chemical potentials correspond to diamond and H_2.

When we relate the enthalpy difference to the bulk solution energies in the preferred octahedral site for carbon, at 0.74 eV [12], and the preferred tetrahedral site for hydrogen, at 0.25 eV [11], we can estimate that the grain boundary will be co-segregated by hydrogen and carbon without separation into hydrogen-free and carbon-free regions. The reason is a relative shift in the chemical potentials of about 0.49 eV in favor of a partial substitution of carbon with hydrogen at the grain boundary.

From this perspective, it is interesting to consider the influence of aluminum, which is known as protective agent against hydrogen embrittlement. The ab initio data we use to estimate the influence of aluminum has been published in [16]. In this publication, there were no grain boundary or bulk solution energies for hydrogen in Fe-Al reported, but the diffusion barriers of hydrogen in carbide-free bainite have been calculated. Since Fe-(Al)-H was chosen as a representative system, we use the reported results as the upper limit for the repulsive interaction of Al with hydrogen in the direct neighborhood. Depending on the Al fraction in the supercell, the shift in the hydrogen solution energy at the transition point along the tetrahedral-tetrahedral path varies and reaches up to 350 meV.

We choose an intermediate value of 150 meV that reflects two aspects: the transition point shift due to Al alloying—at least in fcc Fe—is around twice as high as the shift of the low energy occupied sites solution energies due the Al alloying [8], and even with Al segregating to ferritic grain boundaries [17], we do not assume a local peaking to 50 atomic percent of Al. However, even with 150 meV shift per hydrogen atom, the preference of hydrogen to segregate to the grain boundaries is suppressed.

Besides this thermodynamic picture, we briefly consider possible kinetic scenarios for the substitution of carbon with hydrogen in grain boundaries. While the carbon atoms act as a strengthening element in grain boundaries, as reported in detail in [4], they also increase the local electronic density. Specifically, as reported also in [9], the carbon atom in the grain boundary shows overlapping bands with its adjacent Fe atoms and a spherically symmetric local increase of the charge density distribution (see Figure 3 in [9]). It is possible that the hydrogen atoms in the metallic matrix, which are partially negatively polarised, lead to a local charge redistribution at these spots of increased electronic density, providing access to a substitution of the carbon by hydrogen. We note that this mechanism will require a more detailed quantum mechanical study of the dynamic bonding behavior at this site.

5. Hydride Precipitation near Grain Boundaries

As described in Section 2, the application of a tensile strain can lower the solution enthalpy of interstitial elements and therefore favor segregation. If the material is homogeneously deformed, also the shift of the solution enthalpy is the same everywhere. For a grand canonical ensemble, the hydrogen concentration raises uniformly. In contrast, for a canonical ensemble with a fixed total amount of hydrogen, the concentration distribution is unaffected. This is different for non-uniform deformation, e.g., localized tensile strains near crack tips or dislocations, where the equilibrium hydrogen concentration can strongly increase and eventually lead to localized hydride formation.

As we have worked out recently, solubility limits can also be affected by the proximity of surfaces and interfaces even in the absence of external stresses [18]. At the solubility limit, a secondary phase (hydride) forms, which typically has an elastic mismatch with the parent phase. This induces long-ranged coherency stresses, which tend to suppress the phase separation. If free surfaces are close to the potentially forming precipitate, the stresses can partially be released, and therefore the elastic energy penalty for two-phase situations is reduced. Consequently, there is a preference for phase separation near surfaces, or, in other words, the solubility limit for hydrogen is reduced in these regions.

We have worked out a scale bridging description of these processes, where all relevant material parameters are derived from density functional theory, and the long-ranged elastic distortions are treated on the continuum level. Both descriptions are linked by a statistical mechanics and phase field perspective for the prediction of stress and surface affected solubility limits. A particular finding is that, in the low concentration and low temperature regime, which is most relevant for hydrogen in metals and steel, the solubility limits between bulk and surface differ by a factor

$$s = \exp\left(\frac{\gamma \Delta G_{el}^{bulk}}{kT}\right). \tag{22}$$

This expression is independent of chemical bonding effects and only depends on elastic parameters. It contains the bulk elastic energy

$$\Delta G_{el}^{bulk} = -\frac{E}{1-\nu} \chi^2 \frac{\Omega}{N_0} c_\beta^0, \tag{23}$$

with Young's modulus E, Poisson ratio ν, Vegard coefficient χ and low temperature hydrogen concentration c_β^0 in the precipitate phase. In Equation (22), γ is a dimensionless geometrical factor of order unity, which describes the near surface relaxation (for $0 < \gamma < 1$), which can be obtained e.g., by finite element simulations [19]. We note that this prediction is based in particular on a strain dependent energy including hydrogen, analogous to carbon in Fe as analyzed in Section 2. For a detailed discussion, we refer to [18].

We can use these results to discuss the possibility of precipitation, in particular hydride formation, near grain boundaries. For that, we assume that the grain boundary appears on mesoscopic scales as a layer with a locally different mechanical strength. The inclusion is assumed to have a diagonal eigenstrain relative to the matrix phase, and the elastic constants are taken as equal. A sketch of the geometry is shown in Figure 3.

A spherical precipitate is placed at a certain distance from either a free surface or a grain boundary. If the precipitate is neither close to the free surface nor to the grain boundary, the elastic energy reaches the bulk value, which can be calculated analytically under the given conditions (see [18] for details). For different distances of the precipitate to a free surface or the grain boundary and different elastic properties of the grain boundary, the elastic energy of the system is calculated with the finite element method. Close to the free surface, the elastic energy drops with the distance of the precipitate's center of mass to the surface h (see Figure 4).

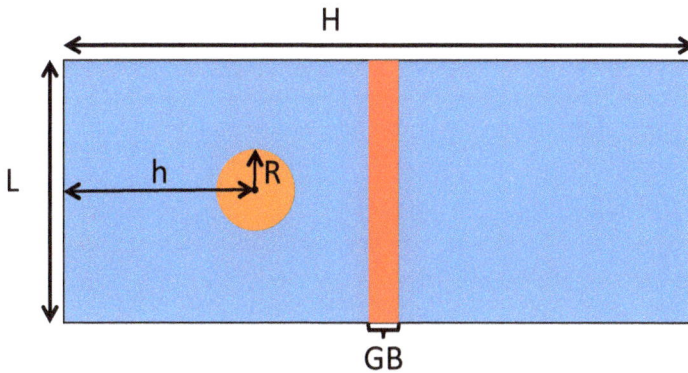

Figure 3. Spherical precipitate in the vicinity of a free surface or grain boundary. The Young's modulus of the grain boundary is effectively either higher or lower than the value in the bulk. The sketch shows a two-dimensional cut through the three-dimensional system.

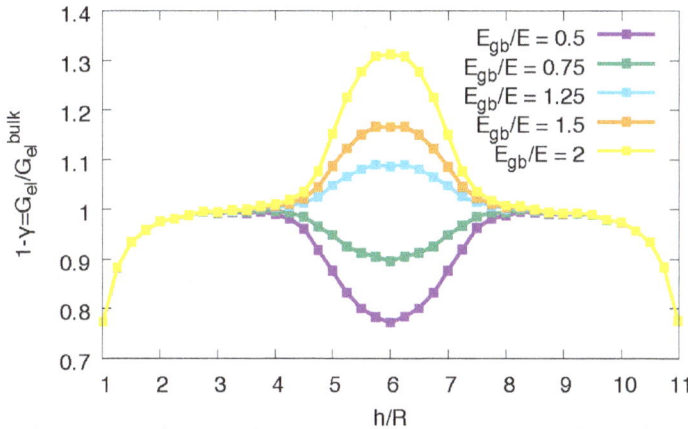

Figure 4. Elastic energy of the two-phase system with a precipitate near a free surface, in the bulk, or close to a hard or soft grain boundary. The energy is shown relative to the bulk energy value. The location of the grain boundary is at $h/R = 6$, and we use a grain boundary thickness $d/R = 0.8$.

When the precipitate approaches the grain boundary, the elastic energy can either increase or decrease, depending on whether the elastic constant E_{gb} in the grain boundary layer is effectively higher or lower than in the bulk. In the limit that E_{gb}/E tends to zero, the grain boundary acts effectively as a free surface, and we recover the same elastic energy relaxation. On the contrary, for $E_{gb} \gg E$, the grain boundary becomes rigid, and we receive the same situation as for nucleation near a fully confined surface, as discussed in detail in [18].

We note that, due to the computational expense, limited datasets of usable mechanical data for grain boundaries are available. In [9], the dependence of the theoretical strength σ_{th} of the $\Sigma 5 (310) [001]$ grain boundary on the hydrogen and carbon population is reported. By varying the carbon and hydrogen content of this grain boundary, σ_{th} varies by 10–50 percent when the empty grain boundary is populated with hydrogen or carbon. In particular, an energetic preference is reported in [9] for a segregation of hydrogen to the carbon saturated grain boundary to replace carbon, and an associated decrease of the theoretical strength by about 20 percent is determined. We note that this value is subject to variations arising due to orientational changes and elastic anisotropy in real samples.

We assume that these effects lead effectively to a local alternation of the ratio E_{gb}/E in the spirit of our continuum picture.

The energy modification $1 - \gamma = G_{el}/G_{el}^{bulk}$ can then be inserted into Equation (22) to obtain the increase or decrease of the solubility limit. Near a stiff grain boundary, hydride formation is less likely.

In conjunction with the role of Al as a protective agent against grain boundary embrittlement discussed in the previous section, the findings in this section suggest the possibility of an increased weakness against hydride formation close to grain boundaries when the segregation of Al to grain boundaries prevents the substitution of carbon by hydrogen. However, we mention that a detailed understanding of the role of Al on carbon population of attractive grain boundary sites is presently lacking to give a more complete estimate here. In particular, the possible configurations of Al that range from a disperse distribution to highly localised concentrations require further investigations on the quantum mechanical scale.

We remark that one can interpret the Boltzmann factor in Equation (22) as a kinetic modification of the energy barrier for a hydride nucleation process. Apart from phase separation triggered by nucleation events, for higher concentrations, also barrierless transformations driven by spinodal decomposition are conceivable. The mixing enthalpy (per unit volume) is given by

$$g_{mix}(c) = \alpha \frac{N_0}{\Omega} c(1-c) \tag{24}$$

as a function of the hydrogen concentration c, using the number of metal atoms N_0 per unit cell of volume Ω and the interaction strength parameter α that can be obtained by ab initio simulations. The dominant configurational contribution to the lattice gas entropy is in the low concentration regime

$$\mu_c = kT \ln c. \tag{25}$$

According to the seminal work by Cahn and Hilliard [20,21], the chemical spinodal is determined by the condition $\mu'(c) = 0$. Asymptotically for $T \to 0$ and for the low concentration branch,

$$c \sim \frac{kT}{2\alpha}. \tag{26}$$

In contrast, the coherent spinodal, which is comparable to the bulk elastic solubility limit, is determined by the condition [20]

$$\mu'(c) + 2\frac{E\chi^2 \Omega}{(1-\nu)N_0} = 0. \tag{27}$$

In the low temperature limit for the low concentration branch, we get

$$c \sim kT \left(2\alpha - 2\frac{E\chi^2 \Omega}{(1-\nu)N_0} \right)^{-1}. \tag{28}$$

Surface spinodal modes have been discussed [22], and they can appear in between the coherent and the surface spinodal in the vicinity of a free surface. From these expressions, we can conclude that, for hydrogen related problems in steel, the required concentrations for spinodal decomposition are typically too high, and phase separation should therefore be initiated by nucleation, as discussed above.

Within this section, we have focused on the influence of mechanical characteristics of grain boundaries on the hydride formation in the vicinity of a grain boundary. Analogously to the thermodynamic description of the influence of elastic relaxation close to free surfaces in [18], the stiffness of the grain boundary can lead to a local decrease or increase of the hydrogen solubility limit, hence favoring or suppressing hydride precipitation from the host matrix.

6. Discussion

We have presented several interlinked scenarios for the continuum level interpretation of ab initio data on hydrogen in steel in the context of hydrogen embrittlement, also in competition with carbon. The connecting motif is the high complexity of the hydrogen embrittlement process at grain boundaries, which requires understanding of all the presented phenomena in a scale-bridging fashion to develop a quantitative connection to macroscale transport and mechanical phenomena.

The connection of ab initio data on the dependence of the solution enthalpy of carbon on the strain state to a continuum scale description supports the discussion of the influence of grain boundaries on hydride formation (or precipitation in the vicinity of grain boundaries in general), also depending on the mechanical properties of the grain boundary. Here, we recognise that, depending on the hydrogen segregation at the grain boundary, a change of the stiffness by about 20 percent can be expected—and this implies a possible competitive embrittlement mechanism either in the vicinity of the grain boundary or in the grain boundary by hydride formation or enrichment. A more conclusive prediction will require more ab initio data, especially since energies can significantly depend on the type of grain boundary, misorientation, etc.

Furthermore, we have described the statistical connection of the ab initio data on the influence of Al on hydrogen solution energies to its local concentration in steel. This also contributes to the discussion of the protective influence of Al against hydrogen embrittlement at grain boundaries. We have introduced an approximative scheme to estimate the energies of grain boundaries for varying levels of co-segregation by hydrogen and carbon, with an accuracy that is comparable to the variation within published ab initio data. We note that, in a carbon saturated ground state of the considered grain boundary, the hydrogen enrichment is driven by a segregation energy. It gives a relative preference to hydrogen instead of carbon by about 100 meV, and we estimate the repulsive effect of Al on H in the grain boundary to be about 150 meV at sufficiently high Al concentrations. From this, a suppressive effect of Al on the hydrogen accumulation at the grain boundary can be expected. For higher Al concentrations, competing phases may form that can also affect the intergranular fracture behavior.

7. Materials and Methods

All numerical approaches for obtaining the simulation data have been obtained using approaches that are explained in detail in the main cited references. In particular, the Vienna ab initio simulation package (VASP, version 5.4.1, Computational Materials Physics, University of Vienna, Vienna, Austria, 2015) has been used for density functional theory calculations [23–25]. Finite element method calculations have been performed with the open source software FreeFEM++ (FreeFEM++, version 3.59) [19].

8. Conclusions

To conclude, we mention that the possibilities to connect available ab initio data sets to continuum descriptions require more data from the ab initio side. Especially for grain boundary hydrogen embrittlement, a comprising study would benefit from data on the dependence of carbon and hydrogen solution energies in at least two different grain boundaries on the presence of aluminum and the strain state. However, these kinds of calculations are computationally expensive, though certain synergy effects are accessible—the theoretical strength can be considered as a byproduct of such simulations. Still, the configurational complexity of the corresponding simulation cells due to the distribution of Al as a substituting element and carbon and hydrogen as interstitials creates a vast set of configurations to be covered. For sure, such data will help to shed light on the complex interplay of different mechanisms, which may contribute to the understanding and prevention of hydrogen embrittlement in the future.

Acknowledgments: This work has been supported by the Collaborative Research Center 761 "Stahl ab initio" of the German Research Foundation. The authors gratefully acknowledge the computing time on the supercomputer JURECA at Forschungszentrum Jülich. The authors thank Tilmann Hickel for many valuable discussions.

Author Contributions: The results in this paper have been worked out together by all authors. The manuscript has been written together by all authors.

Conflicts of Interest: The authors declare no conflict of interest.

References

1. Venezuela, J.; Liu, Q.; Zhang, M.; Zhou, Q.; Andrej, A. A review of hydrogen embrittlement of martensitic advanced high-strength steel. *Corros. Rev.* **2016**, *34*, 153–186.
2. Jemblie, L.; Olden, V.; Akselsen, O.M. A review of cohesive zone modelling as an approach for numerically assessing hydrogen embrittlement of steel structures. *Philos. Trans. R. Soc. A* **2017**, *375*, 20160411.
3. Robertson, I.M.; Sofronis, P.; Nagao, A.; Martin, M.L.; Wang, S.; Gross, D.W.; Nygren, K.E. Hydrogen embrittlement understood. *Metall. Mater. Trans. A* **2015**, *46*, 2323–2341.
4. Hristova, E.; Janisch, R.; Drautz, R.; Hartmaier, A. Solubility of carbon in a-iron under volumetric strain and close to the Σ5(310)[001] grain boundary: Comparison of DFT and empirical potential methods. *Comput. Mater. Sci.* **2011**, *50*, 1088–1096.
5. Hüter, C.; Friák, M.; Weikamp, M.; Neugebauer, J.; Goldenfeld, N.; Svendsen, B.; Spatschek, R. Nonlinear elastic effects in phase field crystal and amplitude equations: Comparison to ab initio simulations of bcc metals and graphene. *Phys. Rev. B* **2016**, *93*, 214105.
6. Von Appen, J.; Dronskowski, R.; Chakrabarty, A.; Hickel, T.; Spatschek, R.; Neugebauer, J. Impact of Mn on the Solution Enthalpy of Hydrogen in Austenitic Fe-Mn Alloys: A First-Principles Study. *J. Comput. Chem.* **2014**, *35*, 2239–2244.
7. Song, E.J.; Bhadeshia, H.K.D.H.; Suh, D.-W. Interaction of aluminium with hydrogen in twinning-induced plasticity steel. *Scr. Mater.* **2014**, *87*, 9–12.
8. Hüter, C.; Dang, S.; Zhang, X.; Glensk, A.; Spatschek, R. Effects of Aluminum on Hydrogen Solubility and Diffusion in Deformed Fe-Mn Alloys *Adv. Mater. Sci Eng.* **2016**. doi:10.1155/2016/4287186.
9. Tahir, A.M.; Janisch, R.; Hartmaier, A. Hydrogen embrittlement of a carbon segregated Sigma 5 (310) [001] symmetrical tilt grain boundary in α Fe. *Mater. Sci. Eng. A* **2014**, *612*, 462–467.
10. Jiang, D.E.; Carter, E.A. Diffusion of interstitial hydrogen into and through bcc Fe from first principles. *Phys. Rev. B* **2004**, *70*, 064102.
11. Du, Y.A.; Ismer, L.; Rogal, J.; Hickel, T.; Neugebauer, J.; Drautz, R. First-principles study on the interaction of H interstitials with grain boundaries in α and γ-Fe. *Phys. Rev. B* **2011**, *84*, 144121.
12. Wang, J.; Janisch, R.; Madsen, G.K.H.; Drautz, R. First-principles study of carbon segregation in bcc iron symmetrical tilt grain boundaries. *Acta Mater.* **2016**, *115*, 259–268.
13. Cak, M.; Sob, M.; Hafner, J. First-principles study of magnetism at grain boundaries in iron and nickel. *Phys. Rev. B* **2008**, *78*, 054418.
14. Bhattacharya, S.K.; Tanaka, S.; Shiihara, Y.; Kohyama, M. Ab initio perspective of the [110] symmetrical tilt grain boundaries in bcc Fe: application of local energy and local stress. *J. Mater. Sci.* **2014**, *49*, 3980–3995.
15. Gao, N.; Fu, C.-C.; Samaras, M.; Schäublin, R.; Victoria, M.; Hoffelner, W. Multiscale modelling of bi-crystal grain boundaries in bcc iron. *J. Nucl. Mater.* **2009**, *385*, 262–267.
16. Li, Y.; Chen, C.; Zhang, F. Al and Si Influences on Hydrogen Embrittlement of Carbide-Free Bainitic Steel. *Adv. Mater. Sci. Eng.* **2013**, 382060.
17. Lejcek, P.; Hofmann, S. Prediction of enthalpy and entropy of grain boundary segregation. *Surf. Interface Anal.* **2002**, *33*, 203–210.
18. Spatschek, R.; Gobbi, G.; Hüter, C.; Chakrabarty, A.; Aydin, U.; Brinckmann, S.; Neugebauer, J. Scale bridging description of coherent phase equilibria in the presence of surfaces and interfaces *Phys. Rev. B* **2016**, *94*, 134106.
19. Hecht, F. New development in FreeFem++. *J. Numer. Math.* **2012**, *20*, 251–266.
20. Cahn, J.W. Coherent Fluctuations and Nucleation in Isotropic Solids. *Acta Metall.* **1962**, *10*, 907–913.
21. Cahn, J.W.; Hilliard, J.E. Free Energy of a Nonuniform System. I. Interfacial Free Energy. *J. Chem. Phys.* **1958**, *28*, 258–267.
22. Tang M.; Karma, A. Surface Modes of Coherent Spinodal Decomposition. *Phys. Rev. Lett.* **2012**, *108*, 265701.
23. Kresse, G.; Hafner, J. Ab initio molecular dynamics for liquid metals. *Phys. Rev. B* **1993**, *47*, 558.

24. Kresse, G.; Furthmüller, J. Efficiency of ab-initio total energy calculations for metals and semiconductors using a plane-wave basis set. *Comput. Mater. Sci.* **1996**, *6*, 15–50.

25. Kresse, G.; Furthmüller, J. Efficient iterative schemes for ab initio total-energy calculations using a plane-wave basis set. *Phys. Rev. B.* **1996**, *54*, 11169.

MDPI

Article

Modeling of Phase Equilibria in Ni-H: Bridging the Atomistic with the Continuum Scale

Dominique Korbmacher [1], Johann von Pezold [1], Steffen Brinckmann [1], Jörg Neugebauer [1], Claas Hüter [2,3] and Robert Spatschek [2,3,*]

[1]   Max-Planck-Institut für Eisenforschung GmbH, D-40237 Düsseldorf, Germany;
      d.korbmacher@mpie.de (D.K.); j.pezold@mpie.de (J.v.P.); s.brinckmann@mpie.de (S.B.);
      j.neugebauer@mpie.de (J.N.)
[2]   Institute for Energy and Climate Research, Forschungszentrum Jülich GmbH, D-52428 Jülich, Germany;
      c.hueter@fz-juelich.de
[3]   Jülich Aachen Research Alliance (JARA), RWTH Aachen University, D-52056 Aachen, Germany
*   Correspondence: r.spatschek@fz-juelich.de; Tel.: +49-2461-61-4470

Received: 29 March 2019; Accepted: 16 April 2018; Published: 18 April 2018

Abstract: In this paper, we present a model which allows bridging the atomistic description of two-phase systems to the continuum level, using Ni-H as a model system. Considering configurational entropy, an attractive hydrogen–hydrogen interaction, mechanical deformations and interfacial effects, we obtained a fully quantitative agreement in the chemical potential, without the need for any additional adjustable parameter. We find that nonlinear elastic effects are crucial for a complete understanding of constant volume phase coexistence, and predict the phase diagram with and without elastic effects.

Keywords: phase equilibrium; Monte Carlo modeling; thermodynamics; elasticity

---

1. Introduction

The understanding of phase equilibria and transitions is essential for the comprehension of many processes in nature as well for a theory guided search for novel materials with superior properties. The modeling of two-phase coexistence in binary alloys is based on the common tangent construction: In equilibrium, the chemical potentials of each species in the two phases of interest have to be equal. As shown in pioneering work by several authors [1–3], elastic effects can significantly modify the phase coexistence behavior, especially in solid phases, e.g., due to density differences.

For a true multi-scale modeling of phase equilibria and transitions in complex materials, ranging from macroscopic dimensions down to the nanoscale, an efficient and accurate matching between the atomistic simulations and formal thermodynamic and continuum concepts is critical. However, despite significant progress, there is still a substantial gap between these two levels. The purpose of the present article is therefore to seamlessly connect the atomistic and continuum scale, illustrated for the Ni-H system. This system is used as a prototype in atomistic simulations for the understanding of hydrogen embrittlement phenomena [4] and rechargeable batteries [5–7]. Mechanical deformations play a crucial role, and thus a careful thermodynamic inspection is essential for a thorough understanding. In previous work [8], we connected ab initio modeling of coherent phase equilibria in the presence of lattice strains and interfacial proximity effects to continuum descriptions. Here, in contrast, the focus is on a matching between Monte Carlo (MC) simulations using empirical potentials and bulk effects on the continuum level. Additionally, we consider interfacial energy contributions, which were neglected in [8].

Experimentally, it is known that metallic nickel absorbs very limited amounts of hydrogen under ordinary conditions, but is known to form a nearly stoichiometric hydride NiH at high hydrogen fugacities. Fukai et al. performed in situ measurements of the lattice parameter of

Ni-H alloys over a wide range of hydrogen partial pressures. This way they constructed a pressure–composition–temperature diagram, encompassing the region of two-phase coexistence [9,10].

One of the central outcomes of the present study is the importance on nonlinear elastic effects in constant volume simulations, as large stresses can arise [11]. Moreover, our approach also provides important fundamental insights into the theory of phase equilibria in coherent solid-state systems, as it elucidates quantitatively the role of different energetic and entropic contributions.

The present work is complementary to atomistic modeling [12] and cluster expansion approaches [13,14], which aim at a description on the atomic level. Cluster expansion approaches are based on an energy description of an Ising-like model of alloys and are powerful methods for predicting phase diagrams and ordering phenomena; long-ranged elastic effects, however, are difficult to capture in such approaches. Therefore, in the present work, we focused on the bridging between a hybrid molecular static and Monte Carlo approach for finding the equilibrium configurations on the one hand and long-wavelength continuum theories on the other hand. This link across the scales is needed to capture in a closed description both microscopic effects related to H–H interactions as well as long-ranged elastic effects due to phase separation. The derived scale bridging energy functional can then serve as a direct input e.g., to phase field simulations. The resulting quantitative scale-bridging is the major outcome of this paper. Step by step, we introduce the various contributions on the continuum scale to match the atomistically determined chemical potential of hydrogen, which is a key quantity in the description of phase coexistence.

2. Methods

The starting point of our study is the determination of the equilibrium spatial distribution of the interstitial H atoms in the metallic matrix, employing grand canonical Monte-Carlo simulations and molecular statics. The simulations are performed within a periodic $10 \times 10 \times 10$ fcc supercell containing 4000 Ni atoms at constant volume (corresponding to the equilibrium volume of pure Ni) and temperature. H atoms are introduced on the octahedral interstitial sites according to the Metropolis algorithm [15]. The energies of the trial steps are determined by relaxing the ionic degrees of freedom—thus fully capturing elastic interactions—using the LAMMPS simulation package [16,17] in conjunction with a modified version [4] of the Ni-H EAM potential by Angelo et al. [18,19]. The effects of temperature are thus only considered to account for the configurational entropy contributions to the free energy, while vibrational contributions are disregarded. This description has previously been used to model hydrogen mediated dislocation–dislocation interaction [4].

The continuum modeling involves the construction of a free energy functional, which is derived step by step in the following section. This partly requires solving continuum mechanics problems, for which we used in particular finite element modeling using ABAQUS.

3. Results

3.1. Monte Carlo Modeling

In the grand canonical simulations, the chemical potential is prescribed, and the number of hydrogen atoms can vary. Depending on the H chemical potential dilute or condensed H distributions are obtained. The condensed hydride precipitates remain coherent and adopt characteristic shapes depending on the bulk H concentration, as shown in Figure 1.

A peculiarity, which motives the present scale bridging investigations, is the functional form of the hydrogen chemical potential μ_H as function of the average H concentration, see dots in Figure 2, which will be discussed in detail below. Classically, we would expect (in a strain free situation) the chemical potential to be constant in the entire two-phase region as a consequence of the common tangent construction. Obviously, this is not the case here, and it is one of the goals of this paper to shed light on this effect.

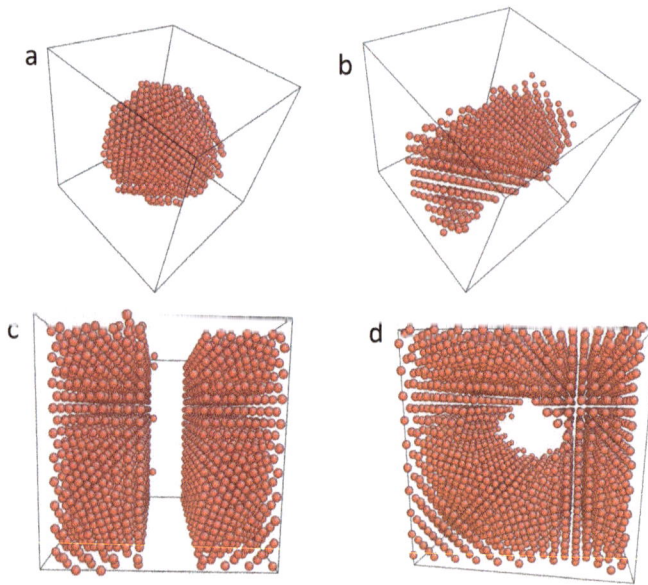

Figure 1. Selected atomistic simulation results for different hydrogen concentrations with fixed volume at $T = 300\,\text{K}$. The Ni matrix is not shown, and the hydrogen atoms are visualized by the dots. The equilibrium shape of the precipitates in a periodic system depends on the H concentrations. First, hydride spheres appear at low concentrations, here for $c = 0.22$ (**a**); followed by a hydride tube (**b**) with $c = 0.23$. For higher saturations (here $c = 0.78$), slabs are observed (**c**). Beyond $c = 0.5$, the situation is inverted and Ni precipitates form inside the hydride matrix. Snapshot (**d**) shows a Ni tube for $c = 0.85$. The last possible shape—a Ni sphere inside a hydride matrix—is not shown.

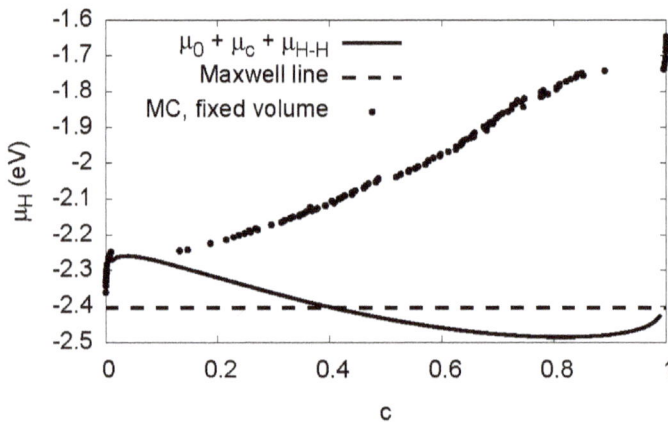

Figure 2. Chemical potential of hydrogen as function of H concentration at $T = 300\,\text{K}$. The S-shaped curve is the single phase prediction, the horizontal dashed line the Maxwell construction. For them, elastic effects are not taken into account, in contrast to the Monte Carlo data, which is for fixed volume (lattice constant of pure Ni). For high concentrations, compressive elastic effects become dominant. Apart from the dilute regions $c \ll 1$ and $1 - c \ll 1$, the Monte Carlo data correspond to two-phase configurations.

3.2. Free Energy Formulation

Our starting point for transferring the atomic scale behavior to the continuum level is the free energy for the single phase material,

$$F = \mu_0 N_H + F_c + F_{H\text{-}H} + F_{el} \tag{1}$$

with the number of hydrogen atoms N_H and the elastic free energy F_{el}, the configurational free energy F_c and the H–H interaction $F_{H\text{-}H}$. μ_0 is the solvation energy needed to insert an isolated hydrogen atom into the (empty) matrix, in contrast to the aforementioned chemical potential μ_H, which also includes mutual interactions, elastic effects, etc.

3.3. Solvation Energy and H–H Interaction

For the formation of a hydride phase, it is essential to include a description of the lattice mediated interaction between the hydrogen atoms [4]. We use the Margules model [5], $F_{H\text{-}H}/N_{Ni} = -\alpha c^2/2 + \beta c^3/3$, where N_{Ni} is the number of nickel atoms and $c = N_H/N_{Ni}$ the (homogeneous) hydrogen concentration; it is normalized to 1 for a crystal where all interstitial octahedral sites are occupied. The parameters are determined from the Monte Carlo calculations by averaging over several random mixtures of fully relaxed homogeneous solutions. We point out that, for this matching, we only consider homogeneous "lattice gas" configurations and not phase separated states, which involve additionally (coherency) elastic and interfacial effects, and which will be parametrized below. These reference calculations are performed under constant pressure conditions, $P = 0$, to suppress (external) stress effects. The resulting curve $F(c)$ is shown in Figure 3. It is fitted by a polynomial $F_{fit}/N_{Ni} = a_0 + a_1 c + a_2 c^2 + a_3 c^3$. Identification with the above parameters therefore uniquely gives $\mu_0 = a_1$, $\alpha = -2a_2$ and $\beta = 3a_3$ (a_0 is an irrelevant zero-point energy). Hence, we obtain $\mu_0 = -2.148\,\text{eV}$ relative to monatomic hydrogen in vacuum, $\alpha = 0.751\,\text{eV}$ and $\beta = 0.355\,\text{eV}$. These numbers differ slightly from those given in Ref. [5] due to a different pair potential cutoff to stabilize the hydride, as this leads to a positive value of the elastic constant C_{44}, see Ref. [4] for details. The positive sign of α favors an increase of the local hydrogen concentration, thus leading to the hydride formation.

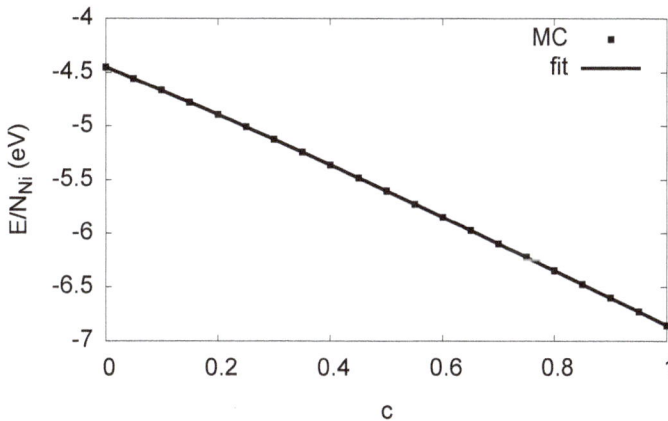

Figure 3. Energy as function of the H concentration for a homogeneous and pressure free system, to determine the solvation energy μ_0 and the H–H interaction. Notice that neither entropic effects are present at $T = 0$, nor elastic or interfacial effects. The Monte Carlo data is averaged over several configurations.

The chemical potential is defined in the usual way,

$$\mu_H = \left(\frac{\partial F}{\partial N_H} \right)_{T,V,N_{Ni}} , \tag{2}$$

and is split into additive contributions analogous to the free energy in Equation (1). We therefore get for the chemical potential contribution by the H–H interaction $\mu_{H\text{-}H} = -ac + bc^2$. It is shown in Figure 4 for both the present fitting parameters and the ones given in Ref. [5], exhibiting only small differences between the predictions.

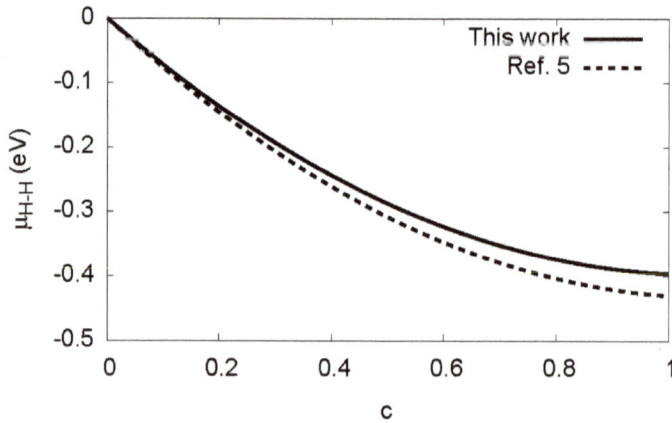

Figure 4. The chemical potential contribution $\mu_{H\text{-}H}$ as function of the hydrogen concentration in a homogeneous system.

3.4. Configurational Entropy

The configurational free energy F_c stems from the different possibilities to occupy the interstitial sites with hydrogen (or the hydride with vacancies), and is therefore given by

$$F_c = k_B T N_{Ni}[(1-c)\ln(1-c) + c \ln c] \tag{3}$$

under the assumption of equal occupation probability for all octahedral sites [20]. In the low concentration regime, it gives the dominant nontrivial contribution, as elastic and interaction effects are negligible there, see Figure 5.

We note that the configurational term is the only temperature dependent one in the present description. To verify this dependence we also computed in the Monte Carlo simulations cases with different temperatures, and the results are shown in Figure 6, showing very good agreement.

3.5. Maxwell Construction

Thus far, the system is described for a single phase state only, and elastic effects are suppressed by a free volume expansion. Due to homogeneity, elastic stresses as well as interfacial contributions do not appear, and the description is therefore complete on this level.

According to the usual thermodynamic picture, a single phase state is unstable in regions where the slope of the chemical potential is negative, $\mu'_H(c) < 0$, and phase separation should occur there. As can be seen in Figure 6, this is the case for a wide concentration regime for low temperatures, whereas for high temperatures the hydrogen solubility limit is significantly larger.

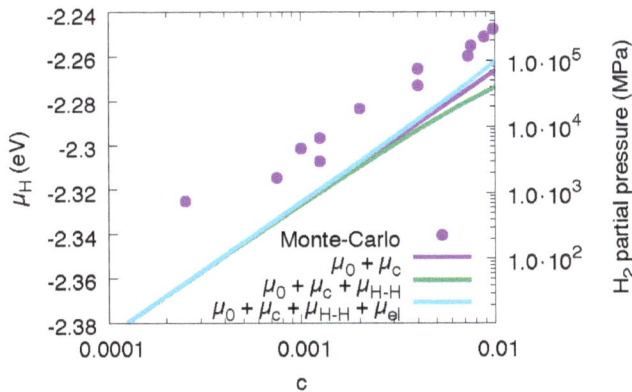

Figure 5. Dilute limit of the chemical potential, which is dominated by the configurational entropy. The data are for $T = 300\,$K and for fixed volume with the equilibrium lattice constant of pure Ni. The chemical potential is (apart from the offset μ_0) dominated by the configurational contribution μ_c, whereas elastic and H–H interaction terms, μ_{el} and $\mu_{H\text{-}H}$ are negligible there. The accuracy of the continuum description in comparison to the Monte Carlo data are finally about 20 meV per atom in the entire concentration range $0 < c < 1$.

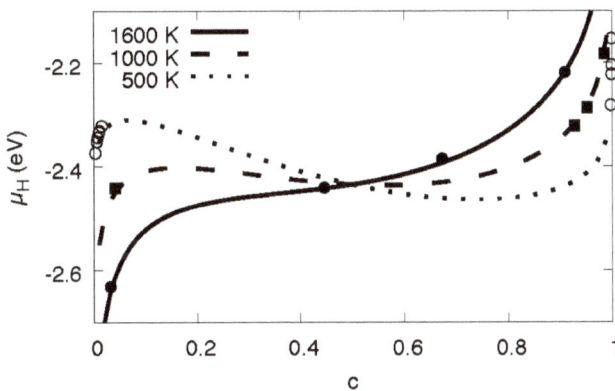

Figure 6. Chemical potential in the single phase state for different temperatures. The curves are the analytical descriptions based on the expression $\mu_0 + \mu_{H\text{-}H} + \mu_c$ using the parameters of the H–H interaction and the solvation energy. The points are the data from the grand canonical Monte Carlo simulations of homogeneous, single-phase states. At low temperatures, the solubility is very low, and therefore the concentration range of the single phase equilibrium states limited to the dilute regions.

Phase coexistence on this level is described by Maxwell's equal area rule, which states that phase separation sets in at the intersection points of a horizontal line with the S-shaped van der Waals loop, cutting it into two equal areas above and below this Maxwell line. This is shown in Figure 2 for $T = 300\,$K. From this, we get the constant chemical potential in the two-phase region, $\mu_M = -2.405\,$eV for $T = 300\,$K. In equilibrium, the system therefore enters the two-phase region at the first intersection of the horizontal Maxwell line with the S-shaped single phase curve. Apparently, this happens already for very low concentrations in the ppm range. Nevertheless, on the following ascending part of the curve $\mu_H(c)$, the system is still metastable. We point out that, according to the principles of the two-phase construction, the dual phase hydrogen chemical potential should be constant and equal to

μ_M until the phase transformation is completed. Obviously, this is not the case for the atomistic data with the fixed volume constraint.

3.6. Elastic Effects

Before entering into a detailed analysis of elastic effects, let us briefly discuss them in relation to the H–H interaction term. The latter is sometimes also introduced as long-range elastic interaction, and we want to stress that we do not double count effects here. The H–H interaction term expresses that an H atom locally deforms the lattice, and therefore makes the placement of another H atom in the vicinity energetically favorable. In a mean field context, this is expressed by the term $\sim -c^2$ in $F_{H\text{-}H}$. If the site occupancy gets higher and the octahedral sites more and more filled, the energy for placing more H atoms into the lattice increases again, as expressed by the $+c^3$ term. Although this consideration makes reference to elastic deformations, this is distinct from the elastic effects considered in the following. To make this point more explicit, we consider the following two situations, for simplicity both with homogeneous hydrogen distributions, i.e., spatially constant concentration c:

First, for fixed pressure $P = 0$, we increase the hydrogen concentration. This leads to a widening of the lattice, as discussed in the following and shown in Figure 7. However, since the system remains stress free on a mesoscopic or macroscopic level, $\sigma_{ij} = 0$, the elastic energy (see Equation (5)) in the continuum mechanics sense remains zero. Nevertheless, the total energy of the system is changed due to the H–H interaction, and this is expressed through $F_{H\text{-}H}$, which is independent of the elastic stress state.

Second, for fixed concentration (and total number of hydrogen atoms), we change the external pressure of the system. Then, the H–H interaction term does not change, but the mesoscopic elastic energy does.

The difference between the H–H interaction term and the explicit elastic term becomes most prominent in two-phase situations. Then, the hydride has a larger equilibrium lattice constant and therefore distorts the surrounding, coherently connected matrix phase. This leads to long-ranged elastic deformations, and the range of these interactions of the order of the precipitate size, which is captured by the mesoscopic elastic energy term introduced below, but not by the H–H interaction term. For further discussion of this separation of microscopic and mesoscopic elastic contributions, we refer to [8]. There, it was shown that the elastic energy due to microscopic deformations around individual H atoms and long ranged strains resulting from the formation of mesoscopic precipitates decompose additively into microscopic and mesoscopic contributions without the appearance of a cross term.

The large deviations between the atomistic data and the continuum model clearly indicate that elastic and interfacial effects are critical and cannot be neglected. We remind that the basis for the Maxwell construction—or here equivalently the common tangent construction—is based on the assumption that the two-phase energy in a macroscopic system decomposes additively into the contributions of the two phases, which do not influence each other. This means, that they are described by free energy densities, which depend only on the local concentration. This condition, however, is violated in the presence of elastic effects. Here, e.g., a density variation in one phase also influences the other, as the global pressure in the system changes.

In the following, we discuss the modeling mainly from the continuum perspective. The technical aspects of the parameter matching between the scales are given in the appendices.

We focus first on the low-temperature regime, where the hydrogen solubility limit is low (see Figure 13), and higher temperatures will be discussed later. In the present regime, the phases are almost stoichiometric, and the chemical potential of the hydrogen in the two-phase region is modified. From a minimization of the total free energy in the two-phase region, we obtain the modified chemical potential (see Appendix A)

$$\mu_H(c) = \mu_M + \mu_{el}(c) + \mu_s(c). \tag{4}$$

In contrast to the Maxwell term μ_M, the elastic and interfacial contributions, μ_{el} and μ_s, respectively, do depend on the concentration, and therefore the chemical potential is no longer constant in the two-phase region, as observed in the Monte Carlo simulations (see Figure 2).

3.6.1. Linear Elastic Effects

The pure nickel and the hydride exhibit a substantial lattice mismatch, and therefore, in the two-phase region, elastic stresses arise. The linear elastic free energy density per unit volume is given by

$$f_{el} = \frac{1}{2} C_{ijkl} (\epsilon_{ij} - \epsilon_{ij}^0)(\epsilon_{kl} - \epsilon_{kl}^0), \tag{5}$$

where $\epsilon_{ij} = (\partial_i u_j + \partial_j u_i)/2$ is the strain derived from the displacements u_i, and the eigenstrain $\epsilon_{ij}^0 = \epsilon_0 \delta_{ij}$ represents the isotropic volume expansion of the hydride [21]. It is concentration dependent, and we extracted it again from $T = 0\,\mathrm{K}$ simulations of homogeneous states with $P = 0$; it is well described by $\epsilon_0(c) = b_1 c + b_2 c^2 + b_3 c^3$ with fitting parameters $b_1 = 0.1027$, $b_2 = -0.0598$ and $b_3 = 0.0189$, going beyond a linear dependence in the spirit of Vegard's law. The eigenstrain is directly related to the concentration dependent lattice constants with $a_{Ni} = 3.520$ and $a_{Ni-H} = 3.738$ via $\epsilon_0(c) = [a(c) - a_{Ni}]/a_{Ni}$, since we use here the relaxed, hydrogen-free nickel as reference configuration (see Figure 7).

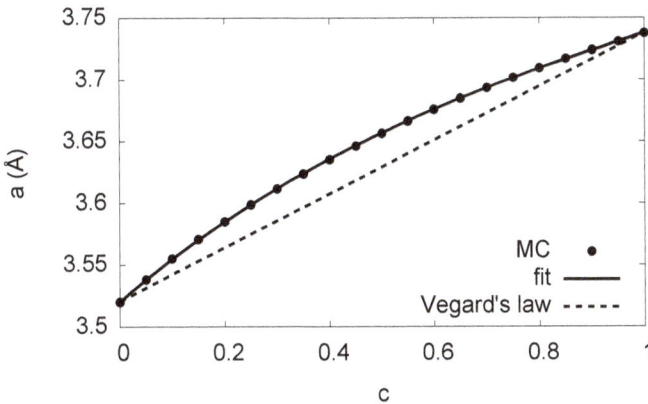

Figure 7. Lattice constant as function of the hydrogen concentration. The Monte Carlo results are obtained from homogeneous free volume simulations, averaged over several configurations. The fit describes the nonlinear concentration dependence, in contrast to a linear interpolation (Vegard's law).

The elastic constants for the pure phases are listed in Table 1. We note that all values are given with respect to Ni as reference configuration; see Appendix B for a discussion of this issue. The determination of the elastic constants from the Monte Carlo simulations is described in detail in Appendix C.

Table 1. Elastic constants of pure nickel and the fully saturated hydride, as obtained from the EAM potential (see Appendix C for details). We note that the relaxed configuration of Ni is used as reference state.

C_{ij}	Ni	Ni-H
C_{11}	250.7 GPa	295.3 GPa
C_{12}	145.5 GPa	197.3 GPa
C_{44}	134.3 GPa	33.5 GPa

To better understand the role of the elastic effects, it is instructive to inspect an analytical isotropic linear elastic model, where we assume for simplicity the elastic constants to be equal in both phases. We use a spherical sample of radius R consisting of nickel, with a hydride inclusion of radius R_H in its center. The radial displacement component is continuous at the coherent interface and vanishes at the outer boundary due to the volume constraint. At low temperatures, the concentration is in good approximation $c = 0$ in Ni and $c = 1$ in the hydride. This gives (see Appendix D)

$$\mu_{\text{lin.el}} = \frac{a_{\text{Ni}}^3 (3\lambda + 2G)[2G + c(3\lambda + 2G)](\Delta\varepsilon_0)^2}{(\lambda + 2G)N_0}, \tag{6}$$

where $N_0 = 4$ is the number of octahedral sites per unit cell, $\Delta\varepsilon_0 = \varepsilon_0(1) - \varepsilon_0(0)$, and λ and G are Lamé coefficient and shear modulus respectively. In this approximation, the elastic contribution in Equation (6) is a linear function of the concentration, and it depends quadratically on the eigenstrain difference $\Delta\varepsilon_0$. Since the material is relaxed for $c = 0$, the elastic effects therefore give only an additive constant for $c = 0$, which is related to the elastic hysteresis due to the internal stresses that arise upon precipitate formation. For the isotropic elastic constants we use the values for nickel and $\lambda = C_{12}$ and $G = (C_{11} - C_{12})/2$ for the solid curve $\mu_M + \mu_{\text{lin.el}}$ in Figure 8.

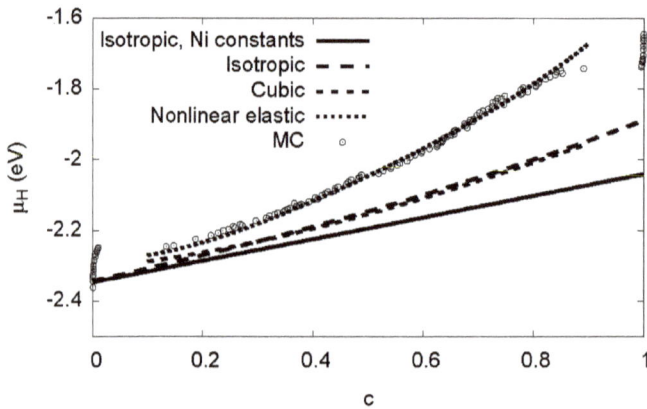

Figure 8. Elastic contribution to the chemical potential in various approximations: The solid curve expresses the isotropic prediction by Equation (6), using equal elastic constants for both phases (we use the values of pure Ni). The long-dashed curve is the same isotropic model, but this time using the different elastic constants of both phases. This prediction is almost identical to a numerically calculated expression based on cubic elasticity, which assumes a spherical inclusion in a cubic box with periodic boundary conditions (short-dashed curve). In contrast to all these linear elastic approximations, the dotted line also considers elastic nonlinearities. Its prediction is very close to the Monte Carlo data.

The calculation can also be done using the elastic constants of the individual phases, leading to the long-dashed curve shown in the same graph (see Appendix D). A numerical finite element solution with ABAQUS for spherical precipitates in a cubic box with fixed volume, taking into account the full cubic elasticity and periodic boundary conditions, leads to the short-dashed curve shown in Figure 8. It is very close to the analytical results. Since we still observe a significant discrepancy to the Monte Carlo data, we conclude that the consideration of linear elasticity is not sufficient to explain the slope of the chemical potential in the two-phase region.

3.6.2. Nonlinear Elastic Effects

The reason for the discrepancy between the atomistic data and the continuum modeling involving linear elasticity is the appearance of large compressive stresses for higher hydrogen concentrations (at fixed volume), and therefore the linear elastic approximation breaks down. Instead, we have to take into account nonlinear elastic effects; they include nonlinearities in the stress-strain relationship, leading effectively to expressions $C_{ijkl}(\{\epsilon_{mn}\})$ in the elastic free energy density. Additionally, geometrical nonlinearities appear due to the volume deformation and the nonlinear strain tensor [11],

$$\epsilon_{ij} = \frac{1}{2}\left(\frac{\partial u_j}{\partial x_i} + \frac{\partial u_i}{\partial x_j} + \frac{\partial u_k}{\partial x_i}\frac{\partial u_k}{\partial x_j}\right). \tag{7}$$

We have estimated that these effects are small in comparison to the change in the constitutive law, and therefore effectively captured them in the modified elastic constants. However, it turns out to be important to carefully use consistently the same reference configuration for the Lagrangian formulation of elasticity for both phases. We note that the chemical potential contains now also a contribution due to the strain dependence of the elastic constants. We mention in passing that we do not observe plastic deformations in the atomistic simulations. From a careful analysis of the Monte Carlo simulations we find that the nonlinear corrections depend only on the volume change and not on shear effects, thus the elastic constants depend only on the trace of the (local) strain tensor, which simplifies the expressions. We therefore write them as series expansion

$$C = C^0 \left[1 + d_1^C \mathrm{tr}(\epsilon - \epsilon^0) + d_2^C [\mathrm{tr}(\epsilon - \epsilon^0)]^2 + \ldots\right] \tag{8}$$

for all elastic constants or combinations of them. From the atomistic simulations we find the coefficients of the nonlinearities for $C_{11} - C_{12}$, C_{44} and the bulk modulus K, see Table 2 and Appendix C. From them, all elastic constants can be expressed using the relation $3K = C_{11} + 2C_{12}$. We note that for the present geometry bulk compression is the dominant effect, and the material becomes stiffer in this regime of negative strain (compression). Including the nonlinear contributions the continuum chemical potential shows now a very satisfactory agreement with the Monte Carlo data, see the dotted line in Figure 8.

Table 2. Coefficients for the nonlinear elastic corrections. If higher order coefficients are missing, the expansion is truncated already at lower order.

Elastic Constant	Material	d_1	d_2	d_3	d_4	d_5
C_{44}	Ni	−3.25	−7.04	–	–	–
$C_{11} - C_{12}$	Ni	−2.01	−12.51	–	–	–
K	Ni	−1.00	0.53	−4.49	2.57	50.03
C_{44}	Ni-H	−22.80	−47.20	–	–	–
$C_{11} - C_{12}$	Ni-H	−9.77	−14.69	–	–	–
K	Ni	−2.25	3.37	11.51	3.55	−51.00

3.7. Interfacial Effects

In a final step, we take into account interfacial effects. As already visible from the agreement of the Monte Carlo data and the nonlinear elastic curve, their contribution is obviously small. In isotropic approximation, the surface energy for a spherical inclusion is given by $F_s = 4\pi\gamma R_H^2$ with the surface energy γ. From planar interface calculations, we obtain $\gamma = 0.11\,\mathrm{Jm}^{-2}$ for (100) interfaces and $\gamma = 0.17\,\mathrm{Jm}^{-2}$ for (111) interfaces. From the fitting of two-phase data, we independently get an excellent agreement using $\gamma = 0.17\,\mathrm{Jm}^{-2}$ for the spherical inclusion.

The magnitude of interfacial terms should be compared to bulk contributions to see their influence and to get an impression on the dimension of critical nuclei. For an order of magnitude estimate, we compare elastic bulk contributions in the approximation in Equation (6) to interfacial contributions for spherical precipitates for small concentrations in Equation (A42), where the influence of interfacial terms is largest. For $c \ll 1$ the elastic and interfacial contributions are comparable for

$$N_H \simeq \frac{4\gamma^3(2G+\lambda)^3 n_0 \pi}{3a_{\mathrm{Ni}}^3(\Delta\epsilon_0)^6 G^3(2G+3\lambda)^3}, \tag{9}$$

which is the case for about 23 H atoms. This indicates that interfacial contributions are indeed small for the present scenario in comparison to the elastic bulk terms.

For a linear isotropic material with equal elastic constants in both phases and a purely dilatational eigenstrain, the elastic energy does not depend on the shape of the inclusion but only on its volume fraction [22]. Although the conditions for this rigorous statement are not exactly fulfilled here, we find only a small dependence of the elastic energy on the precipitate shape in the Monte Carlo data, as exemplarily shown for spherical hydride inclusions in a Nickel matrix and vice versa in Figure 9. Consequently, the shape of the precipitate is determined by the interfacial energy terms alone, see Appendix E for explicit calculations. This means, that for each given average concentration c, which is related to a precipitate size, the equilibrium shape with the minimum interfacial energy appears. For isotropic surface energy we expect spherical inclusions for $c < 4\pi/81 \approx 0.16$, followed by tubes up to $c = 1/\pi \approx 0.32$, and slabs thereafter. The situation is symmetric for $c > 0.5$ with then the hydride being the matrix phase. Thus we expect nickel tubes in the range $1 - 1/\pi < c < 1 - 4\pi/81$ and spheres beyond this concentration. In our Monte Carlo calculations, we indeed find all these structures in the correct ordering (see Figures 1 and 10), but the transition points differ slightly due to anisotropy and nonlinear elastic effects. Based on the observed structures and the calculated interfacial energy we added this contribution to the nonlinear elastic chemical potential in Figure 10. The transitions between different precipitate shapes explain the small discontinuities in the chemical potential, as computed from the Monte Carlo simulations. Apparently, with all the aforementioned effects taken together, we get an excellent agreement with the Monte Carlo data in the entire two-phase region.

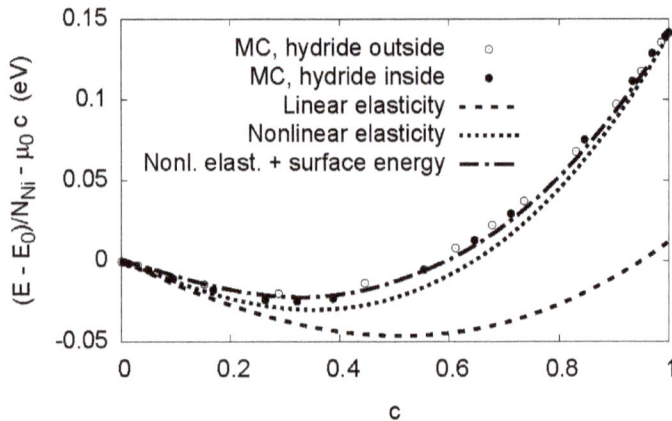

Figure 9. Energy per Ni atom as function of the hydrogen concentration. The atomistic data are calculated at $T = 0\,\mathrm{K}$ without Monte Carlo steps but atomic relaxation for fixed volume. A spherical precipitate is placed in the center to the simulation cell (with periodic boundary conditions). The irrelevant integration constant E_0 for pure Ni is subtracted, as well as the trivial term μ_0, thus only the H–H interaction, elasticity and surface contributions remain. The shape of the precipitate is not important for the elastic energy; we used both Ni-H and Ni inclusions in the atomistic simulation, visualized by open and filled spheres.

Finally, we briefly comment on the role of surface stress. In isotropic approximation, the surface energy for a spherical inclusion is given by $F_s = 4\pi\gamma_0 R_H^2$ with the surface energy γ_0. The surface stress contribution is $F_\beta = 4\pi R_H^2 \beta(\epsilon_{\theta\theta} + \epsilon_{\phi\phi})$ in spherical coordinates, with the scalar surface stress coefficient β [23,24]. We obtain—in the same approximation as for Equation (6)—due to the additional pressure difference between the precipitate and the matrix $\Delta\sigma_{rr} = 2\beta/R_H$ also a contribution from the elastic bulk energy ΔF_{el} [25]

$$F_\beta + \Delta F_{el} = \frac{8}{3}\pi\frac{3\lambda + 2\mu}{\lambda + 2\mu}\beta\Delta\epsilon_0 \, R_H^2, \tag{10}$$

where we have neglected a small destabilizing term proportional to β^2 for $|\beta| \ll 3K\Delta\epsilon_0 R_H$ (we have estimated from the atomistic data β to be of the order $-1\,\text{Jm}^{-2}$, thus this condition is fulfilled), and also assumed $c \ll 1$, since spherical precipitates appear for small concentrations only, as discussed below. Consequently, these terms have the same radius scaling as the surface energy term $\sim R_H^2$, and despite their partial origin from a bulk energy, they appear effectively as surface term; therefore, we treat them via a renormalized interfacial energy density γ.

With this, the parametrization of the continuum model is completed, and it is used in the following for predicting hydrogen partial pressures, volume relaxation effects, and for the prediction of phase diagrams.

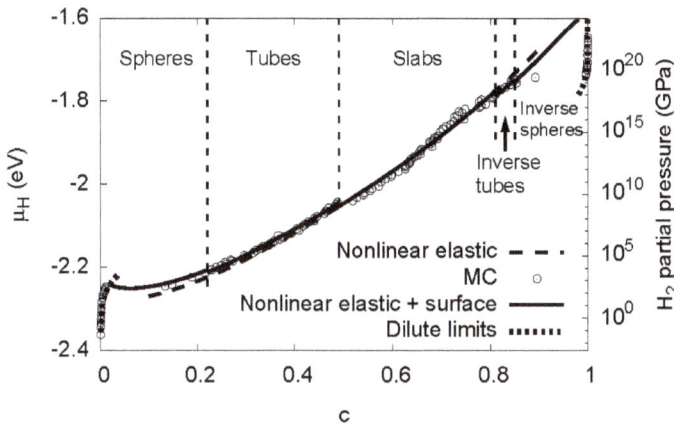

Figure 10. The chemical potential of hydrogen at $T = 300\,\text{K}$. The dots are the data from the Monte Carlo simulations. The chemical potential including nonlinear elastic effects is very close to the Monte Carlo data, and, together with interfacial effects, the agreement is excellent. The contribution from the interfacial terms is largest for low concentrations, as there the precipitates are small and therefore have the largest surface-to-volume ratio. The dotted curves near $c = 0$ and $c = 1$ are the analytical predictions for the single phase dilute limits. The H_2 partial pressure is based on an ideal gas model, which is not realistic for higher pressures and only serves for illustrational purposes.

3.8. High Concentration Limit

We can also correctly describe the limit of high hydrogen concentrations, where almost all octahedral sites are filled with H. For $1 - c \ll 1$, the system is again in a single phase state, with a dilute distribution of vacancies, thus the steep divergency of the chemical potential is due to the configurational degrees of freedom. Nevertheless, the central difference to the dilute limit $c \ll 1$ is that the material is here severely under (nonlinear) stress (as we focus on fixed volume situations, where pure Ni is stress free, hence stresses are maximized for $c = 1$), thus elastic effects play a major role here. In contrast to the phase separated region we need precise knowledge of the concentration dependence of the elastic constants, their nonlinearities and the eigenstrain. Consideration of all these effects leads to the good agreement between the Monte Carlo data and the continuum picture, as shown in Figure 10.

3.9. Conversion to Partial Pressures

To relate the chemical potentials to experimentally accessible quantities, it is useful to translate them to hydrogen (H_2) partial pressures. In the following, we use a subscript H_2 to discriminate between quantities related to the gaseous hydrogen molecules and the monatomic hydrogen in solid solution. From the ideal gas equation and the Maxwell relation

$$\left(\frac{\partial \mu_{H_2}}{\partial p_{H_2}}\right)_{T,N_{H_2}} = \left(\frac{\partial V}{\partial N_{H_2}}\right)_{T,p_{H_2}} = \frac{k_B T}{p_{H_2}} \tag{11}$$

we obtain by integration the chemical potential of H_2 up to an additive term $\mu^0_{H_2}$,

$$\mu_{H_2}(T, p_{H_2}) = \mu^0_{H_2}(T) + k_B T \ln \frac{p_{H_2}}{p_{ref}} \tag{12}$$

with an arbitrary reference pressure p_{ref}. From the equilibrium between H_2 and dissociated hydrogen follows

$$\mu_{H_2} = 2\mu_H. \tag{13}$$

This leads to

$$p_{H_2} = p_{ref} \exp\left(\frac{2\mu_H - \mu^0_{H_2}(T)}{k_B T}\right). \tag{14}$$

We therefore need an expression for $\mu^0_{H_2}(T)$ to link the chemical potential to a hydrogen partial pressure.

In the dilute limit $c \ll 1$, the chemical potential is (see also Figure 5)

$$\mu_H = \mu_0 + k_B T \ln c. \tag{15}$$

In combination with Equations (12) and (13), we therefore get Sievert's law

$$c = \left(\frac{p_{H_2}}{p_{ref}}\right)^{1/2} \exp\left(-\frac{\mu_0 - \frac{1}{2}\mu^0_{H_2}(T)}{k_B T}\right). \tag{16}$$

This is in agreement with experimental findings [26] for the solubility of H_2 in Ni

$$c = 2c_0 \left(\frac{p_{H_2}}{p_0}\right)^{1/2} \exp\left[-\frac{1485.167\,\text{K}}{T} + 3.9881\right], \tag{17}$$

expressed in terms of the quantities used here, with $c_0 = 5.8694 \cdot 10^{-7}$ and $p_0 = 133.3224\,\text{Pa}$ (the prefactor 2 stems from the dissociation of a H_2 molecule in two H atoms). This allows to identify an expression for the offset potential $\mu^0_{H_2}(T)$

$$\begin{aligned}
\mu^0_{H_2}(T) &= 2\mu_0 - k_B \cdot 2970.334\,\text{K} + 7.9762\,k_B T \\
&= -4.552\,\text{eV} + T \cdot 6.873 \cdot 10^{-4}\,\text{eV}\,\text{K}^{-1},
\end{aligned} \tag{18}$$

where we have chosen $p_{ref} = p_0/(4c_0^2)$. With these identifications, Equation (14) is the desired relation, which is shown in Figure 10. Apparently, the H_2 partial pressures needed for the higher concentrations are extremely high and not reachable in experiments under normal conditions, making computer simulations particularly useful there. In addition, we note that for such high pressures the used ideal gas description of H_2 breaks down. On the other hand, the H_2 partial pressures needed for formation of a hydride are much lower for free volume situations, which will be discussed below (see Figure 11). We can therefore conclude that volumetric constraints, or—more generally—compressive stresses can suppress the formation of hydrides, which can be responsible for the embrittlement of the metal.

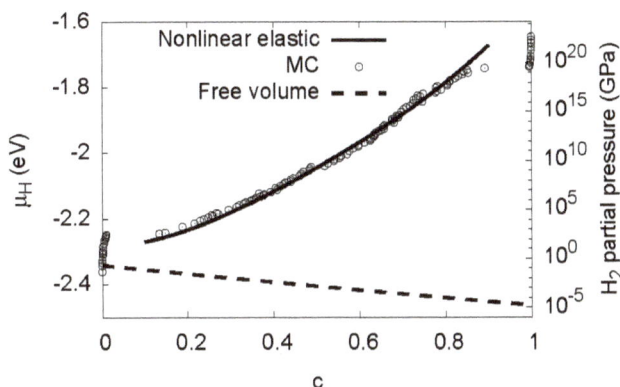

Figure 11. Comparison of the chemical potential for fixed volume and in a stress-free situation, $P = 0$, $T = 300$ K. The volume constraint leads to a positive slope of the chemical potential in the two-phase region, thus stabilizing phase separation of given chemical potential. In contrast, two phase states are unstable for given μ_H, as shown here using the continuum prediction.

3.10. Volume Relaxation

Instead of a fixed volume constraint, one may also consider a situation with free expansion, $P = 0$. This does not mean that elastic effects disappear, because in the two-phase region internal coherency stresses still arise. Thus, the same elastic barrier appears for the first nucleation of the precipitate, but thereafter the elastic contribution to the chemical potential decreases with concentration, see Figure 11. Since the stress effects are much lower, elastic nonlinearities are negligible here, and the curve is calculated in linear isotropic approximation. The qualitatively different behavior shows the important role of external boundary conditions. In contrast to scenarios with free expansion, the volume constraint stabilizes two-phase equilibria. This behavior can intuitively be understood as follows: When a precipitate is inserted into the matrix, coherency stresses arise, which increase the elastic energy. If the precipitate grows and finally fills the whole sample, the freely expanding material is homogeneous and therefore stress free. This implies that the elastic energy decays with the hydride volume fraction, and therefore the elastic contribution to the chemical potential appears with negative sign. Therefore, we get a hydrogen chemical potential with a negative slope in the two-phase region, which corresponds to an unstable situation. We have verified this prediction using Monte Carlo simulations for fixed and free volume at $T = 300$ K, see Figure 12.

We note that volume constraints can therefore stabilize precipitates. In a simulation, this is useful for probing situations with phase coexistence, which would otherwise not appear, as either the system is homogeneously in the dilute state or fully saturated with H for given chemical potential. This is expressed here by a negative slope of the chemical potential of H. The present physically motivated approach is therefore an alternative to bias potentials [12] to probe configurations with phase coexistence.

3.11. Phase Diagram

Based on the temperature dependence of F_c, we can predict the entire bulk phase diagram without elastic effects via the Maxwell construction, see Figure 13. We note that this phase diagram is derived solely from $T = 0$ K data for the H–H interaction. To verify it also for higher temperatures, we have compared the analytical expression for the chemical potential to the single-phase data from grand-canonical Monte Carlo simulations with free volume relaxation, where elastic and interfacial effects do not appear, and find an excellent agreement (see Figure 6). We note that the purpose of the present work is to match a continuum description of Ni-H to Monte Carlo simulations based on an empirical interatomic potential. The agreement between the two theoretical approaches is excellent. Compared to experimental values, however, the description is less suitable for high temperatures, and therefore it is not accurate in this regime. The reason is—apart from potential limitations of the interatomic potential—the use of the (lattice) Monte Carlo molecular statics description, which neglects in particular phonon excitations.

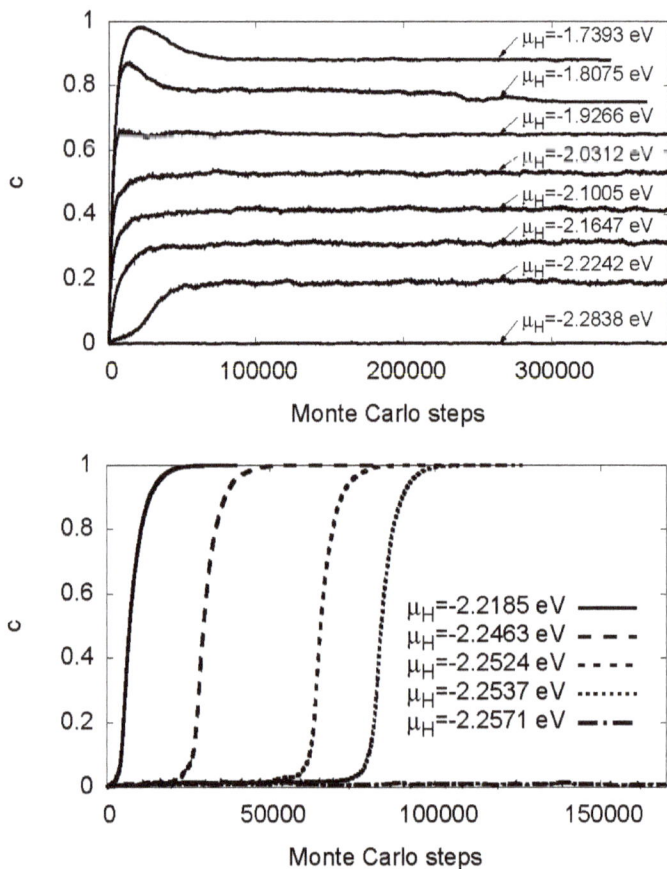

Figure 12. Evolution of the hydrogen concentration as function of the number of Monte Carlo steps for $T = 300\,\mathrm{K}$. Grand-canonical simulations with given chemical potential μ_H are used. (**Top**) Fixed volume using the equilibrium lattice constant of Ni. Due to the positive slope of the chemical potential in the two-phase region (solid curve and data points in Figure 11) stable two phase states with arbitrary average concentrations can be obtained in thermodynamic equilibrium. (**Bottom**) For free volume, the miscibility gap is inaccessible for the given chemical potential in the grand canonical simulations, and instead the equilibrium state is almost free of hydrogen or almost fully saturated, as the solubilities are low at low temperatures (see phase diagram Figure 13). The reason for this behavior (which is accompanied by a hysteresis) is the negative slope of the chemical potential in the two-phase region (see Figure 11), which implies that these configurations are unstable. All simulations are started as a hydrogen free system, and formation of the hydride therefore demands to overcome a nucleation barrier.

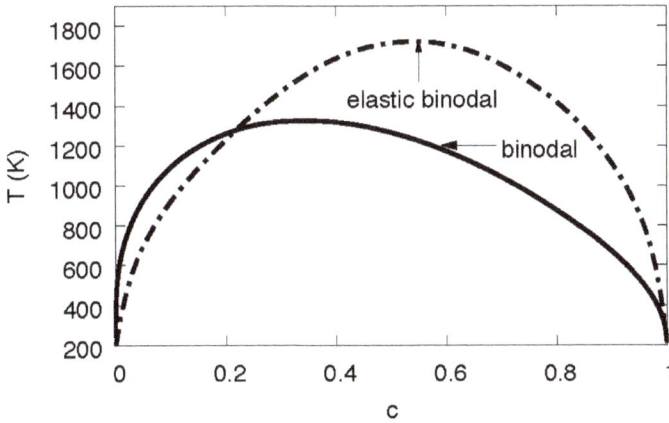

Figure 13. Phase diagram of the Ni-H systems, based on parameters extracted from Monte Carlo simulations. The solid line is the binodal without consideration of elastic effects, whereas the dash-dotted line is the same with elastic effects for fixed volume.

An application that we can readily extract from the above results is the phase diagram taking into account elastic effects. In particular, we consider the situation of fixed volume and determine the two-phase region by minimization of the total free energy. For the sake of simplicity, we assume a slab geometry with the phase normal in [100] direction (see Figure 14), which is the most relevant pattern in the medium concentration regime, as discussed above in Figure 10. In contrast to other geometries, homogeneity implies the constancy of concentrations in each phase. We neglect the elastic nonlinearities, which allows for a straightforward solution of the underlying one-dimensional elastic problem. As the elastic constants were determined mainly for $c = 0$ and $c = 1$, we assume additionally that they interpolate linearly for arbitrary concentrations $0 < c < 1$. This is mainly relevant for higher temperatures, where the appearing phases have larger solubilities for hydrogen or vacancies, thus they appear with intermediate concentrations. As will be discussed below, it turns out that the concentration dependence of the elastic constants is not important, and therefore this approximation is legitimate.

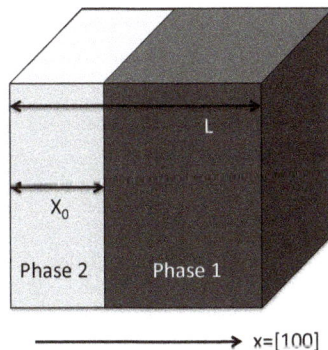

Figure 14. Sketch of the hydride phase (light grey) forming as a slab in the nickel matrix with fixed volume. Due to translational invariance, we can assume the hydride phase to be located in the left part of the cube, starting at $x = 0$.

We inspect the case of a constant total volume which is stress free for pure nickel, i.e., the lattice constant for a homogeneous system is a_{Ni}. Consequently, for the two-phase system with cubic symmetry, the only non-vanishing displacement component is u_x, which is a linear function of x in each phase, $u_x^{(i)} = a_i x + b$ (the phases are enumerated by i). Hence, the only nontrivial strain component is $\epsilon_{xx}^{(i)} = a_i$; stresses follow from Hooke's law with cubic symmetry,

$$\sigma_{xx}^{(i)} = C_{11}(c_i)(a_i - \epsilon_0(c_i)) - 2C_{12}(c_i)\epsilon_0(c_i), \tag{19}$$

$$\sigma_{yy}^{(i)} = \sigma_{zz}^{(i)} = C_{12}(c_i)[a_i - 2\epsilon_0(c_i)] - C_{11}(c_i)\epsilon_0(c_i). \tag{20}$$

Because the concentration is homogeneous in each phase for this effectively one-dimensional situation, the linear bulk elastic equations are automatically fulfilled by the above ansatz. The coefficients a_1, b_1, a_2, b_2 are determined by the boundary and matching conditions, $u_x^{(2)}(x = 0) = 0$, $u_x^{(1)}(x = L) = 0$, $u_x^{(1)}(x_0) = u_x^{(2)}(x_0)$ and $\sigma_{xx}^{(1)}(x_0) = \sigma_{xx}^{(?)}(x_0)$. The elastic energy density is in each phase

$$f_{\text{el}}^{(i)} = 2C_{12}(c_i)\epsilon_0(c_i)[-2a_i + 3\epsilon_0(c_i)] + C_{11}(c_i)[a_i^2 - 2a_i\epsilon_0(c_i) + 3\epsilon_0(c_i)^2]. \tag{21}$$

From that, we can calculate the elastic energy per unit area,

$$F_{\text{el}} = f_{\text{el}}^{(1)}(L - x_0) + f_{\text{el}}^{(2)} x_0. \tag{22}$$

The total free energy is then $F = \mu_0 N_H + F_c + F_{\text{H-H}} + F_{\text{el}}$. Additionally, the concentrations c_1 and c_2 in the two phases are related to the (fixed) average concentration c via the lever rule,

$$Lc = (L - x_0)c_1 + x_0 c_2. \tag{23}$$

We can therefore investigate F/L as function of c_1, c_2 for given values of T, c, and the phase fractions are determined by the lever rule. The concentration domain can be restricted to $c_1 < c < c_2$, and the free energy is minimized with respect to c_1 and c_2 in equilibrium. Whenever the minimum is located inside this domain, the system is in a two-phase state, whereas the minimum is located on the border $c_1 = c$ or $c_2 = c$ for a single phase state. By performing this minimization in the entire c, T plane one obtains the elastic phase diagram.

Already at this level we see the tremendous influence of the elastic effects in Figure 13, with an enlargement of the two-phase region towards higher temperatures and the hydrogen rich side. This unusual behavior—one would intuitively expect the suppression of phase separation since coherency stresses are energetically unfavorable—is due to the fact that the eigenstrain is a concave function of the concentration, $\epsilon_0''(c) < 0$ (see Figure 7), thus volumetric deviations from Vegard's law play the dominant role here.

To investigate this peculiarity in more detail, we look at the elastic energy depending on the phase concentrations. We start with a situation which differs from the true Ni-H problem. Instead, we assume a linear dependence of the eigenstrain on concentration (see Figure 7) (Vegard's law). Here, we find that single-phase states have the lowest elastic energy in a wide region of the phase diagram and are therefore energetically favorable. Thus, phase separation is suppressed by elastic effects in this fictitious situation, leading to an increased H solubility limit, and this is shown in Figure 15. In contrast, for the true concentration dependence of the lattice constant on the H concentration in Figure 7, we obtain a reduced solubility limit of hydrogen in the nickel matrix. As mentioned above, this is a counterintuitive result, as the appearance of phase separation should usually increase the elastic energy and therefore make this state thermodynamically unfavorable, in contrast to our observation. Here, the situation is opposite: Single-phase states are more expensive from point of view of elastic energy, and the hydrogen solubility limit is therefore reduced. The explanation is that the single-phase case has a higher average eigenstrain, and therefore the mechanical volume work is higher than for the phase separated case for the anticipated fixed volume ensemble.

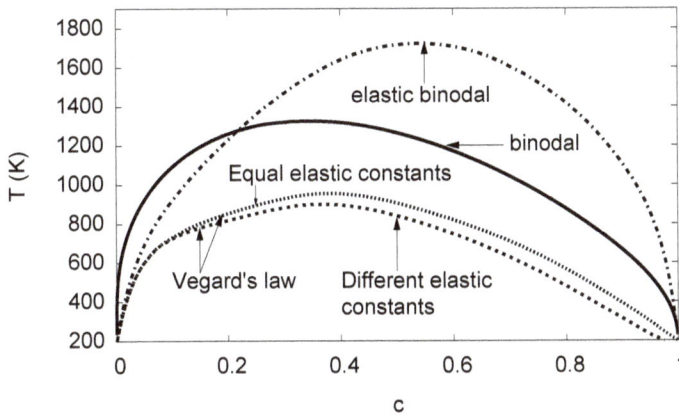

Figure 15. The phase diagram depending on the eigenstrain is shown. The binodal and elastic binodal curves are the same as in Figure 13. The other two curves show the tremendous influence of the concentration dependence of the eigenstrain or lattice constant. If, instead of the concave dependence $\varepsilon_0''(c) < 0$, a linear relationship is assumed (Vegard's law), the solubility of hydrogen is drastically enhanced. In contrast, the concentration dependence of the elastic constants plays only a minor role: The case of different elastic constants anticipates a linear interpolation of the elastic constants between the values for Ni and Ni-H, whereas, for equal constants, both phases are assumed to have the same elastic constants (using values for Ni).

As a result, we find that the $\varepsilon_0(c)$ dependence has a tremendous influence on the phase diagram. This effect is much more pronounced that the concentration dependence of the elastic constant. If we assume, instead of the linear interpolation between the values for Ni and Ni-H, just concentration independent constants using the values of Ni, we find only a small shift of the binodal (see Figure 15). We can therefore conclude that the concentration dependence of the lattice constant is the most relevant quantity for the influence of elastic effects on hydrogen solubility limits.

4. Discussion and Conclusions

Understanding the thermodynamics of phase separation across the scales is a key for deriving macroscopic material properties from fundamental microscopic descriptions. Often, such a transfer is not straightforward, but, with careful descriptions, it is possible to obtain not only macroscopic parameters but also describe energy functionals in agreement with the underlying behavior on atomic scales. We demonstrated such an approach for the Ni-H system with a special focus not only on classical thermodynamics but also strong mechanical and interfacial effects.

In detail, we have seamlessly matched the results of hybrid molecular static and Monte-Carlo simulations in a binary system with full consideration of elastic effects to a continuum description for large scale simulations. The accurate identification of configurational, interaction, nonlinear elastic and surface effects allows for a thorough understanding, which will be equally useful and applicable to other mechanisms or material systems. Presently, the description is most useful for the low temperature regime below the Debye temperature. The consideration of vibrational degrees of freedom and defect formation, as well as the consideration of more complex phase diagrams will be the subject of future investigations. It should be pointed out that the accuracy of the predictions is limited by the atomic scale description, to which the continuum energy functional is tailored. The inclusion of the effects mentioned before will presumably require the consideration of additional energy terms and variables such as defect concentration.

The determination of phase diagrams demonstrates the opportunities arising from the transfer of atomistic data to the mesoscale: The quantitative determination of the free energy allows for example to construct the entire phase diagram in Figure 13 computationally much more efficiently than using atomistic simulations which require large system sizes and long relaxation times, especially in the room temperature regime. Moreover, the description serves as an accurate basis for mesoscale simulation methods such as phase field [27].

Acknowledgments: This work was supported by the DFG Collaborative Research Center 761 *Steel ab initio*. The authors gratefully acknowledge the computing time on the supercomputer JURECA at Forschungszentrum Jülich.

Author Contributions: All authors contributed equally to this work.

Conflicts of Interest: The authors declare no conflict of interest.

Abbreviations

The following abbreviation is used in this manuscript:

MC   Monte Carlo

Appendix A. Thermodynamic Framework

To understand the influence of elastic effects on the phase diagram and the chemical potential, we start with considering the free energy of the system. We use the following notations: N_{Ni} is the total number of nickel atoms, N_{Ni}^d is the number of Ni atoms in the dilute phase, $N_{Ni}^h = N_{Ni} - N_{Ni}^d$ is the number of Ni atoms in the hydride, and n_0 is the number of nickel atoms per unit cell (4 for fcc). Similarly, N_H is the total number of hydrogen atoms, with N_H^d of them being in the dilute phase and $N_H^h = N_H - N_H^d$ in the hydride. In the fully saturated hydride, we have N_0 hydrogen atoms per unit cell (here, $N_0 = 4$), which corresponds to the normalization $c = 1$. Thus, we define the hydrogen concentration in the dilute phase $c^d = N_H^d n_0 / (N_{Ni}^d N_0)$, and in the hydride, $c^h = N_H^h n_0 / (N_{Ni}^h N_0)$. The average concentration is $c = N_H n_0 / (N_{Ni} N_0)$.

The free energy of the system is

$$F = N_{Ni}^d f^d(c^d) + N_{Ni}^h f^h(c^h) + F_{el}(N_{Ni}^d, N_{Ni}^h, c^d, c^h). \tag{A1}$$

Here, f^d is the free energy density (per host atom) of the dilute phase, and f^h is the same for the hydride. The elastic energy depends on the volume fractions and the concentrations in each phase. Here, we already suppressed the degree of freedom of the shape of the hydride inclusion and assume that the free energy is already minimized with respect to it. Notice that we consider here only situations with fixed global strain (i.e., fixed volume), therefore the free energy is the appropriate thermodynamic functional.

As a further approximation of the elastic energy for low temperatures, we take into account that the dilute phase is almost free of hydrogen ($N_H^d \approx 0$), whereas the hydride is almost fully saturated ($N_H^h \approx N_{Ni}^h N_0 / n_0$). This implies $N_H^h \approx N_H$. Therefore, we obtain the simplification $F_{el} \approx N_{Ni} f_{el}(c)$ with the intensive elastic energy density f_{el}.

For the determination of the two-phase equilibrium, we proceed in the usual way: The free energy has to be minimized with respect to the phase fractions and the partitioning. Therefore, we get

$$\left(\frac{\partial F}{\partial N_H^d} \right)_{N_H, N_{Ni}, N_{Ni}^d} = \mu^d(c^d) - \mu^h(c^h) = 0, \tag{A2}$$

where we defined the chemical potentials

$$\mu^{d/h} = \frac{n_0}{N_0} \frac{d}{dc^{d/h}} f^{d/h}(c^{d/h}). \tag{A3}$$

The equality of grand potentials becomes

$$\left(\frac{\partial F}{\partial N_{Ni}^d}\right)_{N_H, N_{Ni}, N_H^d} = f^d(c^d) - c^d \mu^d - f^h(c^h) + c^h \mu^h = 0. \tag{A4}$$

Notice that the elastic terms do not appear in the common tangent construction in the framework of the above approximations, since the concentration is assumed to be fixed in this approximation, which holds here for low temperatures.

Next, we calculate the chemical potential of the hydrogen in the two-phase region. It is given by

$$
\begin{aligned}
\mu_H &= \left(\frac{dF}{dN_H}\right)_{N_{Ni}} = \left(\frac{\partial F}{\partial N_H}\right)_{N_{Ni}, N_{Ni}^d, N_H^d} + \left(\frac{\partial F}{\partial N_{Ni}^d}\right)_{N_{Ni}, N_H, N_H^d} \left(\frac{dN_{Ni}^d}{dN_H}\right)_{N_{Ni}} \\
&+ \left(\frac{\partial F}{\partial N_H^d}\right)_{N_{Ni}, N_H, N_{Ni}^d} \left(\frac{dN_H^d}{dN_H}\right)_{N_{Ni}} \\
&= \left(\frac{\partial F}{\partial N_H}\right)_{N_{Ni}, N_{Ni}^d, N_H^d},
\end{aligned} \tag{A5}
$$

since the last two terms vanish by the equilibrium conditions in Equations (A2) and (A4). Here, we obtain

$$\mu_H = \mu^h + \frac{n_0}{N_0}\left(\frac{\partial f_{el}}{\partial c}\right), \tag{A6}$$

which corresponds to Equation (4) in the main text. Interfacial effects can be considered in a similar way.

Appendix B. Reference State

For the elastic problem, we describe the material in the Lagrangian reference frame of the stress-free nickel. The eigenstrain of the hydride is therefore $\epsilon_{ij}^0 = [a_{Ni-H} - a_{Ni}]/a_{Ni}\delta_{ij}$; notice that also for the hydride the (eigen-)strain is defined relative to the lattice constant of the relaxed Ni. To make this point more transparent, we denote all quantities with Ni-H as reference state by a tilde, e.g., a dilatational strain by $\tilde{\epsilon}_{ij} = [a - a_{Ni-H}]/a_{Ni-H}$, thus apart from an additive eigenstrain contribution $\epsilon_{ij} = \tilde{\epsilon}_{ij} \times a_{Ni-H}/a_{Ni}$. Since stresses are defined as force per area in the reference configuration, we have the transformation rule $\sigma_{ij} = \tilde{\sigma}_{ij} \times (a_{Ni-H}/a_{Ni})^2$. From Hooke's law, $\sigma_{ij} = C_{ijkl}\epsilon_{kl}$ and $\tilde{\sigma}_{ij} = \tilde{C}_{ijkl}\tilde{\epsilon}_{kl}$, we therefore conclude $C_{ijkl} = \tilde{C}_{ijkl} \times a_{Ni-H}/a_{Ni}$. This means that that the elastic constants of the hydride relative to nickel as reference state appear by a factor a_{Ni-H}/a_{Ni} higher than their nominal values for Ni-H reference configuration.

Appendix C. Elastic Constants

Here, we briefly explain the extraction of the elastic constants from the Monte Carlo simulations. Notice that in the absence of vibrational contributions the elastic constants are temperature independent and are therefore calculated at $T = 0$ K. In all cases, the relaxed equilibrium configuration of pure Ni is taken as reference state, see Appendix B.

Appendix C.1. Bulk Modulus

To get the bulk modulus K for the phases, we strain the system isotropically with $\epsilon_{xx} = \epsilon_{yy} = \epsilon_{zz} = \epsilon$. The resulting energies are represented in Figure A1. In linear elasticity, the energy increases quadratically with the strain,

$$f_{el} = \frac{9}{2}K\epsilon^2, \tag{A7}$$

and we determine the bulk modulus K from the curvature of the measured energy–volume curves. Hence, we get for pure Nickel $K_{Ni} = 180.5$ GPa and for the fully saturated hydride $K_{Ni-H} = 230.0$ GPa.

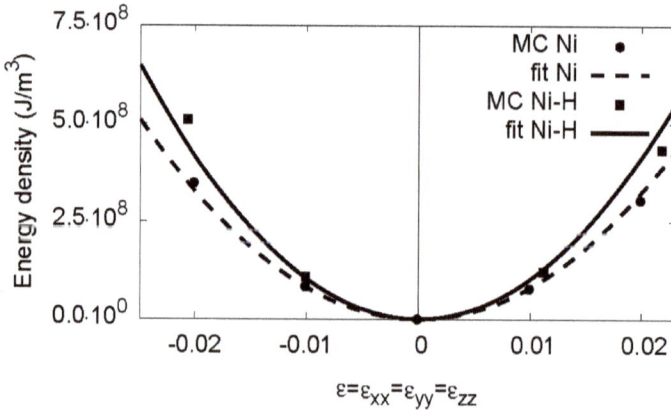

Figure A1. Energy density as function of the external hydrostatic strain ϵ. With the relation in Equation (A7), we get the bulk modulus K in quadratic approximation. The atomistic data are taken from simulations at $T = 0$ K. For high strains deviations from the linear elastic response can be seen.

Appendix C.2. Monoclinic Distortion

The monoclinic distortion describes shear effects to get the elastic constant C_{44} both for Ni and the hydride. One gets the translational movement of an atom from the original location $\vec{r} = \vec{r}(x, y, z)$ to the new position $\vec{r}' = \vec{r}'(x', y', z')$,

$$\vec{r}' = \begin{pmatrix} x + \alpha y/2 \\ \alpha x/2 + y \\ [1 + \alpha^2/(1 - \alpha^2)]\, z \end{pmatrix} \tag{A8}$$

with a deformation parameter α. With the displacement $\vec{u} = \vec{r}' - \vec{r}$ and the linear strain tensor $\epsilon_{ij} = (\partial_i u_j + \partial_j u_i)/2$, we obtain

$$\epsilon_{xy} = \frac{\alpha}{2} \quad \text{and} \quad \epsilon_{zz} = \frac{\alpha^2}{1 - \alpha^2} = O(\alpha^2), \tag{A9}$$

whereas all other strains are zero and the quadratic term in α can be neglected in a linear approximation. Thus, the elastic energy is

$$f_{el} = \frac{1}{2}\sigma_{ij}\epsilon_{ij} = \frac{1}{2}C_{44}\alpha^2. \tag{A10}$$

Figure A2 shows the energy density of both phases depending on the strain ϵ_{xy} for the monoclinic distortion, and the elastic constant C_{44} is obtained from the curvature at the origin. We obtain a value of $C_{44} = 134.3$ GPa for Nickel and $C_{44} = 33.5$ GPa for the hydride. Notice that for pure shear the quadratic energy–deformation dependence holds up to several percent of strain.

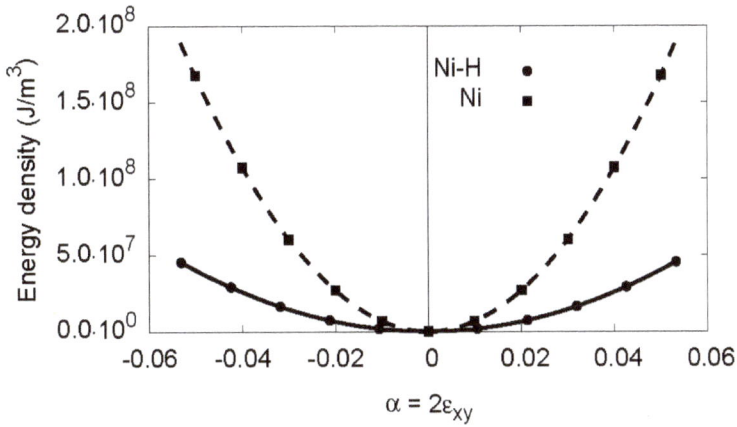

Figure A2. The results for the monoclinic distortion are shown. The atomistic data is fitted via a quadratic ansatz.

Appendix C.3. Orthorhombic Distortion

The orthorhombic distortion delivers the result for the term $C_{11} - C_{12}$ for both phases. It is described by

$$\vec{r'} = \begin{pmatrix} (1+\alpha)x \\ (1-\alpha)y \\ [1+\alpha^2/(1-\alpha^2)]\,z \end{pmatrix}, \tag{A11}$$

hence

$$\epsilon_{xx} = \alpha, \quad \epsilon_{yy} = -\alpha \quad \text{and} \quad \epsilon_{zz} = \frac{\alpha^2}{1-\alpha^2}, \tag{A12}$$

where ϵ_{zz} can be neglected again with the same argument as above. Consequently, the elastic energy is

$$f = \frac{1}{2}\sigma_{ij}\epsilon_{ij} = (C_{11} - C_{12})\alpha^2. \tag{A13}$$

The elastic constants C_{11} and C_{12} are

$$C_{11} = K + \frac{2}{3}(C_{11} - C_{12}), \tag{A14}$$

$$C_{12} = K - \frac{1}{3}(C_{11} - C_{12}). \tag{A15}$$

From the quadratic fit in Figure A3, we get for nickel $C_{11} = 250.7\,\text{GPa}$, $C_{12} = 145.5\,\text{GPa}$, and for the hydride $C_{11} = 295.3\,\text{GPa}$, $C_{12} = 197.3\,\text{GPa}$.

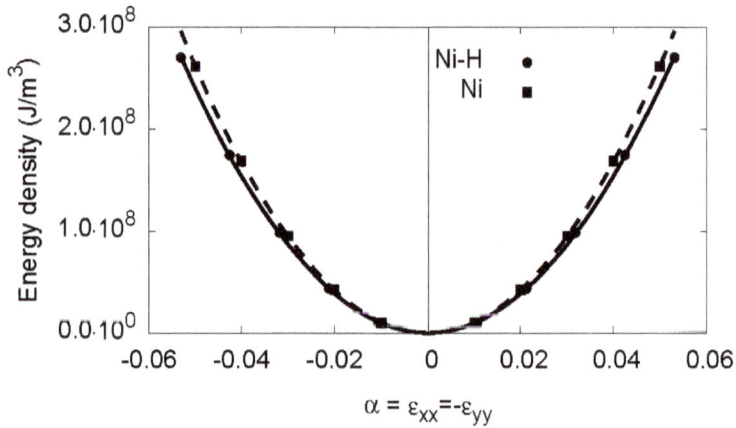

Figure A3. The results for the orthorhombic distortion are shown. The atomistic data are fitted via a quadratic ansatz.

Appendix C.4. Nonlinear Elastic Coefficients

To demonstrate the extraction of the elastic nonlinearities, we inspect the mechanical response of pure Ni under bulk compression or expansion, as shown in Figure A4.

For larger strains, we see deviations from the linear response, i.e., a quadratic energy–strain relationship. For tensile stresses the material appears effectively softer, whereas it is stiffer under compression. In the strain regime shown in the plot, the elastic response is well described by

$$f_{el} = \frac{9}{2} K \epsilon^2 \left(1 + d_1^K \mathrm{tr}\epsilon + d_2^K (\mathrm{tr}\epsilon)^2 + d_3^K (\mathrm{tr}\epsilon)^3 + \dots \right), \tag{A16}$$

in the spirit of Equation (8). The coefficients d_i^K are given in Table 2.

For pure shear during a monoclinic transformation, the energy is well described by a quadratic strain dependence, as shown before in Figure A2, and the same holds for the volume preserving orthorhombic transformation (see Figure A3). This demonstrates that the nonlinearities depend only on the trace of the strain tensor. We have performed monoclinic and orthorhombic transformations at different lattice units (i.e., bulk compression or expansion), to extract the coefficients in the spirit of Equation (8). This is exemplarily shown in Figure A5 for the monoclinic deformation of pure Ni. This allows to extract the elastic constants, here e.g., C_{44} as function of the trace of the strain tensor. This dependence is shown in Figure A6. The resulting data for both phases are then described by Equation (8) with the coefficients in Table 2. The corresponding plot for the orthorhombic deformation is shown in Figure A7.

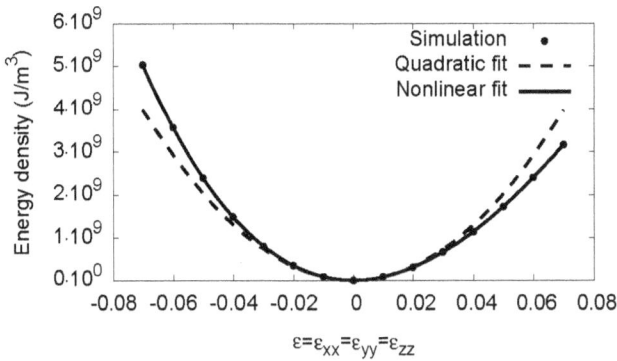

Figure A4. The free energy density as a function of the external strain ϵ for the pure Nickel. The quadratic ansatz is only true for small strains, whereas for higher strain we have to use a nonlinear function to fit the Monte Carlo data.

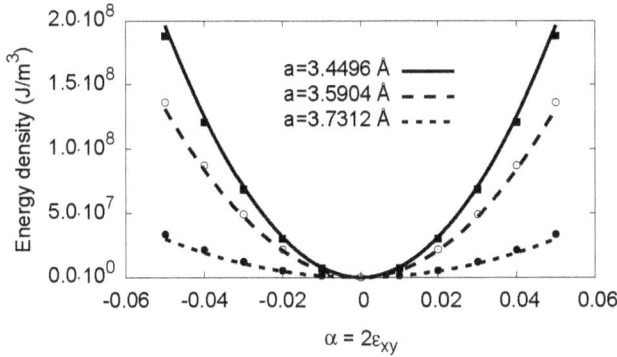

Figure A5. Monoclinic deformation of pure nickel for different bulk lattice constants. The fits (curves) are parabolic, and the data (points) are well described without further nonlinear corrections, in agreement with the statement that corrections depend only on $\mathrm{tr}\epsilon$. The latter is contained in the lattice constant dependence of the curvatures, leading to strain-dependent elastic constants $C_{44}(\mathrm{tr}\epsilon)$. The energy of the bulk compression is subtracted in the plots.

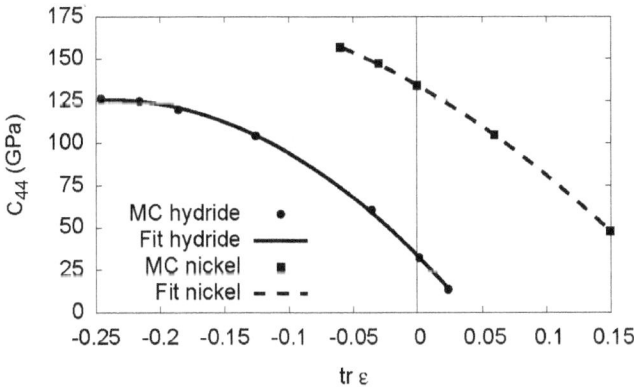

Figure A6. The elastic constant C_{44} as function of the trace of the strain $\mathrm{tr}\epsilon$ for Nickel and the hydride is shown. For $\mathrm{tr}\epsilon = 0$, we obtain the values: $C_{44,Ni} = 134.3$ GPa and $C_{44,H} = 33.5$ GPa.

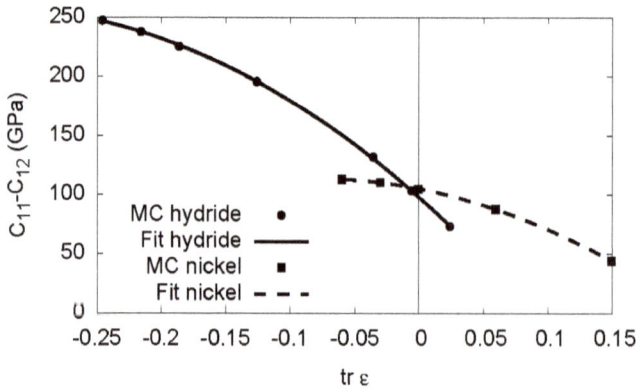

Figure A7. The difference $C_{11} - C_{12}$ depending on the trace of strain tr ϵ for Nickel and the hydride is shown.

Appendix D. The Isotropic Model

We assume that the material is isotropic, and that a spherical nucleus of the hydride with radius R_H forms within the dilute matrix. To simplify the geometry, we assume that the system is also spherical with radius R, thus the entire problem is rotational invariant. Hence, the displacement field depends only on the radius coordinate and has only a radial component. We note that we treat here only the linear elastic case, where corrections due to higher order terms in the strain as introduced in Equation (8) are not considered. Nevertheless, we note that the nonlinear case can also be treated in a similar way. However, for going beyond the one-dimensional case for an inclusion inside a cubic cell, we use finite element simulations instead.

In the outer dilute phase, the displacement is given by

$$u_r^d = a^d r + b^d / r^2, \tag{A17}$$

with constants a^d and b^d. This ansatz satisfies the linear elastic bulk equations. In the hydride (inner phase), we use similarly

$$u_r^h = a^h r. \tag{A18}$$

The strains are given in spherical coordinates by

$$\epsilon_{rr}^{(i)} = u_r^{(i)'}, \qquad \epsilon_{\theta\theta}^{(i)} = \epsilon_{\phi\phi}^{(i)} = u_r^{(i)} / r, \tag{A19}$$

and all non-diagonal terms vanish. The stresses in the dilute phase are

$$\sigma_{ij}^d = \lambda^d \epsilon_{kk}^d \delta_{ij} + 2 G^d \epsilon_{ij}^d, \tag{A20}$$

and in the hydride

$$\sigma_{ij}^h = \lambda^h (\epsilon_{kk}^h - 3\Delta\epsilon_0)\delta_{ij} + 2 G^h (\epsilon_{ij}^h - \Delta\epsilon_0\delta_{ij}) \tag{A21}$$

with the eigenstrain $\Delta\epsilon_0\delta_{ij}$.

At the interface coherency and stress balance demands

$$u_r^d(R_H) = u_r^h(R_H), \qquad \sigma_{rr}^d(R_H) = \sigma_{rr}^h(R_H). \tag{A22}$$

At the outer boundary, the fixed volume condition implies

$$u_r^d(R) = 0. \tag{A23}$$

These are altogether three conditions to determine the constants a^d, b^d and a^h. The elastic energy densities (per unit volume) are

$$f_{el}^d = \frac{1}{2}\lambda^d \epsilon_{kk}^{d\,2} + G^d \epsilon_{ij}^{d\,2} \tag{A24}$$

and

$$f_{el}^h = \frac{1}{2}\lambda^h (\epsilon_{kk}^h - 3\Delta\epsilon_0)^2 + G^h (\epsilon_{ij}^h - \Delta\epsilon_0 \delta_{ij})^2. \tag{A25}$$

The total free energy is

$$F_{el} = 4\pi \int_0^{R_H} r^2 f_{el}^h \, dr + 4\pi \int_{R_H}^R r^2 f_{el}^d \, dr, \tag{A26}$$

which is a lengthy expression. Using the conversions

$$\frac{4}{3}\pi R^3 = \frac{N_{Ni}}{n_0} a^3, \qquad c = \left(\frac{R_H}{R}\right)^3 \tag{A27}$$

with the lattice unit a_{Ni}, we can calculate the elastic contribution to the chemical potential

$$\mu_{el} = \frac{n_0}{N_0} \left(\frac{\partial f_{el}}{\partial c}\right). \tag{A28}$$

In the general case, it becomes

$$\begin{aligned}
\mu_{el} = \ & (3a_{Ni}^3 (\Delta\epsilon_0)^2 (3\lambda^h + 2G^h)(c^2(3\lambda^d + 2G^d) \times (3\lambda^d - 3\lambda^h + 2G^d - 2G^h) \\
& + 4G^d (3\lambda^h + 4G^d + 2G^h) + 2c(3\lambda^d + 2G^d)(3\lambda^h + 4G^d + 2G^h))) \\
& / [2(3\lambda^h + 4G^d + c(3\lambda^d - 3\lambda^h + 2G^d - 2G^h) + 2G^h)^2 N_0].
\end{aligned} \tag{A29}$$

In particular for equal elastic constants ($\lambda^d = \lambda^h = \lambda$, $G^h = G^d = G$)

$$\mu_{el} = \frac{a_{Ni}^3 (\Delta\epsilon_0)^2 (3\lambda + 2G)(2G + c(3\lambda + 2G))}{(\lambda + 2G)N_0}. \tag{A30}$$

Appendix E. Interfacial Effects

The shape of the precipitate forming in the two-phase region is mainly determined by interfacial energy. For isotropic surface energy, a spherical inclusion minimizes the interfacial energy for small radii $R_H \ll L$, with L being the edge length of the enclosing box with periodic boundary conditions. However, for $R_H \sim L$, the formation of other interface contours is energetically favorable, since we use periodic boundary conditions in the atomistic and FEM simulations.

To find the different ranges and the limiting concentrations at which the forms change, we have to compute the minimum surface energy. The surface energy is defined as

$$F_s = \gamma S \tag{A31}$$

where S defines the size of the interfaces and γ the surface energy coefficient.

First, we look for the spherical inclusion of the hydride with the radius R_H. In this case, the surface energy is

$$F_s = \gamma S = \gamma 4\pi R_H^2. \tag{A32}$$

The concentration (volume fraction) is given as

$$c = \frac{4\pi R_H^3}{3L^3}, \tag{A33}$$

so the surface energy is

$$F_s(c) = \gamma L^2 (36c^2\pi)^{1/3}. \tag{A34}$$

The surface energy for a hydride tube with radius r and length L is

$$F_s = \gamma 2\pi r L. \tag{A35}$$

with the respective concentration

$$c = \frac{\pi r^2 L}{L^3} \tag{A36}$$

we obtain

$$F_s(c) = 2\gamma L^2 (c\pi)^{1/2} \tag{A37}$$

In the case of slabs, the surface energy is independent of the concentration c:

$$F_s = 2\gamma L^2. \tag{A38}$$

The cases with $c > 1/2$ are analogous, with the role of the matrix and the hydride being exchanged. Figure A8 shows the surface energy depending on the concentration for the respective case. Up to $c = 4\pi/81 \approx 0.16$, it should be a spherical inclusion of the hydride in the nickel matrix. For $4\pi/81 \leq c \leq 1/\pi$, the inclusion of the hydride should be a tube followed by plates up to $c = 1 - 1/\pi$. The analytical results predict a symmetric behavior for concentrations $c > 0.5$, so for concentrations $1 - 1/\pi \leq c \leq 1 - 4\pi/81$ there should be a tube and for $c > 1 - 4\pi/81$ we expect a nickel sphere.fPhaseDiagram3

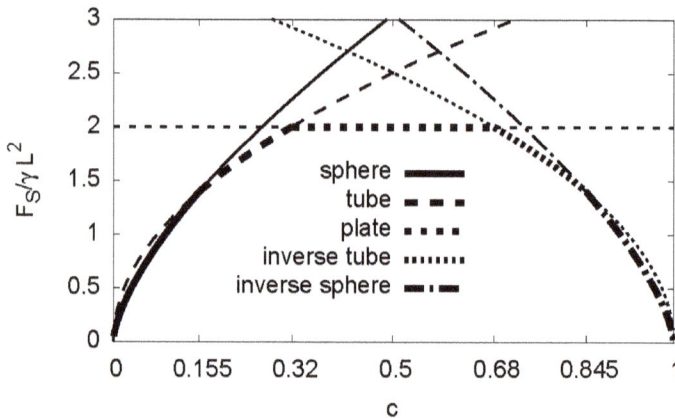

Figure A8. Surface energy depending on the concentration for different geometric cases. The thick line segments indicate the lowest energy configurations.

With the calculated surface energy F_s, we compute now the chemical potential with $\mu_s = dF_s/dN_H$. Then, we have to rewrite the surface energy in an expression which depends on the number of hydrogen atoms N_H. For the first case, the hydride sphere inclusion, the radius can be expressed as

$$R_H = a_{Ni} \left(\frac{3N_H}{N_0 4\pi} \right)^{1/3}, \tag{A39}$$

so

$$F_s(N_H) = a_{Ni}^2 \gamma \pi^{1/3} \left(\frac{6N_H}{N_0} \right)^{2/3}. \tag{A40}$$

The chemical potential then becomes

$$\mu_s(N) = \frac{2 \cdot 2^{2/3} a_{\mathrm{Ni}}^2 \gamma \, (\pi/3)^{1/3}}{N_0 \, (N_{\mathrm{H}}/N_0)^{1/3}}, \tag{A41}$$

and finally with $c = N_{\mathrm{H}} n_0/(N_{\mathrm{Ni}} N_0)$

$$\mu_s(c) = \frac{2 \cdot 2^{2/3} a_{\mathrm{Ni}}^2 \gamma \, (\pi/3)^{1/3}}{N_0 \, (c N_{\mathrm{Ni}}/n_0)^{1/3}}. \tag{A42}$$

Similarly, for tubes, we get

$$\mu_S = \frac{a_{\mathrm{Ni}}^2 \sqrt{\pi} \gamma}{\sqrt{c \, (N_{\mathrm{Ni}}/n_0)^{2/3} N_0}}, \tag{A43}$$

for slabs

$$\mu_S = 0, \tag{A44}$$

for inverse tubes

$$\mu_S = -\frac{a_{\mathrm{Ni}}^2 \sqrt{\pi} \gamma}{N_0 \sqrt{(1-c) \, (N_{\mathrm{Ni}}/n_0)^{2/3}}}, \tag{A45}$$

and for inverse spheres

$$\mu_S = -\frac{4 a_{\mathrm{Ni}}^2 \pi^{1/3} \gamma}{N_0 \, [6(1-c) N_{\mathrm{Ni}}/n_0]^{1/3}}. \tag{A46}$$

References

1. Williams, R.O. The calculation of coherent phase equilibria. *Calphad* **1984**, *8*, 1–14. [CrossRef]
2. Cahn, J.W.; Larché, F. A simple model for coherent equilibrium. *Acta Metall.* **1984**, *32*, 1915–1923. [CrossRef]
3. Johnson, W.C.; Vorhees, P.W. Phase equilibrium in two-phase coherent solids. *Metall. Trans. A* **1987**, *18*, 1213–1228. [CrossRef]
4. Von Pezold, J.; Lymperakis, L.; Neugebauer, J. Hydrogen-enhanced local plasticity at dilute bulk H concentrations: The role of H–H interactions and the formation of local hydrides. *Acta Mater.* **2011**, *59*, 2969–2980. [CrossRef]
5. Haftbaradaran, H.; Song, J.; Curtin, W.A.; Gao, H. Continuum and atomistic models of strongly coupled diffusion, stress, and solute concentration. *J. Power Sour.* **2011**, *196*, 361–370. [CrossRef]
6. Hüter, C.; Fu, S.; Finsterbusch, M.; Figgemeier, E.; Wells, L.; Spatschek, R. Electrode-Electrolyte Interface Stability in Solid State Electrolyte Systems: Influence of Coating Thickness under Varying Residual Stresses. *AIMS Mater. Sci.* **2016**, *4*, 867. [CrossRef]
7. Markus, I.M.; Prill, M.; Dang, S.; Markus, T.; Spatschek, R.; Singheiser, L. High temperature investigation of electrochemical lithium insertion into Li$_4$Ti$_5$O$_{12}$. *Phys. Chem. Chem. Phys.* **2016**, *18*, 31640. [CrossRef]
8. Spatschek, R.; Gobbi, G.; Hüter, C.; Chakrabarty, A.; Aydin, U.; Brinckmann, S.; Neugebauer, J. Scale bridging description of coherent phase equilibria in the presence of surfaces and interfaces. *Phys. Rev. B* **2016**, *94*, 134106. [CrossRef]
9. Shizuku, Y.; Yamamoto, S.; Fukai, Y. Phase diagram of the Ni-H system at high hydrogen pressures. *J. Alloys Comp.* **2002**, *336*, 159. [CrossRef]
10. Fukai, Y.; Yamatomo, S.; Harada, S.; Kanazawa, M. The phase diagram of the Ni-H system revisited. *J. Alloys Comp.* **2004**, *372*, L4–L5. [CrossRef]
11. Hüter, C.; Friák, M.; Weikamp, W.; Neugebauer, J.; Goldenfeld, N.; Svendsen, B.; Spatschek, R. Nonlinear elastic effects in phase field crystal and amplitude equations: Comparison to ab initio simulations of bcc metals and graphene. *Phys. Rev. B* **2016**, *93*, 214105. [CrossRef]
12. Sadigh, B.; Erhart, P. Calculation of excess free energies of precipitates via direct thermodynamic integration across phase boundaries. *Phys. Rev. B* **2012**, *86*, 134204. [CrossRef]

13. Laks, D.B.; Ferreira, L.G.; Froyen, S.; Zunger, A. Efficient cluster expansion for substitutional systems. *Phys. Rev. B* **1992**, *46*, 12587. [CrossRef]

14. Kurta, R.P.; Bugaev, V.N.; Ortiz, A.D. Long-Wavelength Elastic Interactions in Complex Crystals. *Phys. Rev. Lett.* **2010**, *104*, 085502. [CrossRef]

15. Metropolis, N.; Rosenbluth, A.W.; Rosenbluth, M.N.; Teller, A.H. Equation of State Calculations by Fast Computing Machines. *J. Chem. Phys* **1953**, *21*, 1087–1092. [CrossRef]

16. Plimpton, S.J. Fast Parallel Algorithms for Short-Range Molecular Dynamics. *J. Comp. Phys.* **1995**, *117*, 1–19. [CrossRef]

17. LAMMPS Molecular Dynamics Simulator: http://lammps.sandia.gov (accessed on 17 April 2018).

18. Angelo, J.E.; Moody, N.R.; Baskes, M.I. Trapping of hydrogen to lattice-defects in nickel. *Model. Simul. Mater. Sci. Engr.* **1995**, *3*, 289. [CrossRef]

19. Baskes, M.I.; Sha, X.W.; Angelo, J.E.; Moody, N.R. Trapping of hydrogen to lattice defects in nickel. *Model. Simul. Mater. Sci. Eng.* **1997**, *5*, 651–652. [CrossRef]

20. Haasen, P. *Physical Metallurgy*; Cambridge University Press: Cambridge, UK, 2003.

21. Landau, L.D.; Lifshitz, E.M. *Theory of Elasticity*; Pergamon Press: Oxford, UK, 1987.

22. Mura, T. *Micromechanics of Defects in Solids*; Springer: Berlin/Heidelberg, Germany, 1987.

23. Marchenko, V.I. Theory of the equilibrium shape of crystals. *Zh. Eksp. Teor. Fiz. (USSR)* **1981**, *81*, 1141.

24. Fischer, F.D.; Waitz, T.; Vollath, D.; Simha, N.K. On the role of surface energy and surface stress in phase-transforming nanoparticles. *Progress Mater. Sci.* **2008**, *53*, 481–527. [CrossRef]

25. Cahn, J.W.; Larché, F. Surface Stress and Chemical Equilibrium of Small Crystals. II. Solid. Particles Embedded in a Solid Matrix. *Acta Metall.* **1982**, *30*, 51–56. [CrossRef]

26. Armbruster, M.H. The Solubility of Hydrogen at Low Pressure in Iron, Nickel and Certain Steels at 400 to 600°. *J. Am. Chem. Soc.* **1943**, *65*, 1043–1054. [CrossRef]

27. Brener, E.A.; Marchenko, V. I.; Spatschek, R. Influence of strain on the kinetics of phase transitions in solids. *Phys. Rev. E* **2007**, *75*, 041604. [CrossRef]

metals

MDPI

Article

Ab Initio Guided Low Temperature Synthesis Strategy for Smooth Face–Centred Cubic FeMn Thin Films

Friederike Herrig [1,*], **Denis Music** [1], **Bernhard Völker** [1,2], **Marcus Hans** [1], **Peter J. Pöllmann** [1], **Anna L. Ravensburg** [1] **and Jochen M. Schneider** [1]

[1] Materials Chemistry, RWTH Aachen University, Kopernikusstr. 10, 52074 Aachen, Germany; music@mch.rwth-aachen.de (D.M.); b.voelker@mpie.de (B.V.); hans@mch.rwth-aachen.de (M.H.); poellmann@mch.rwth-aachen.de (P.J.P.); ravensburg@mch.rwth-aachen.de (A.L.R.); schneider@mch.rwth-aachen.de (J.M.S.)

[2] Max-Planck-Institut für Eisenforschung GmbH, Max-Planck-Straße 1, 40237 Düsseldorf, Germany

* Correspondence: herrig@mch.rwth-aachen.de; Tel.: +49-241-80-25968

Received: 30 April 2018; Accepted: 24 May 2018; Published: 26 May 2018

Abstract: The sputter deposition of FeMn thin films with thicknesses in the range of hundred nanometres and beyond requires relatively high growth temperatures for the formation of the face-centred cubic (*fcc*) phase, which results in high thin film roughness. A low temperature synthesis strategy, based on local epitaxial growth of a 100 nm thick *fcc* FeMn film as well as a Cu nucleation layer on an α-Al_2O_3 substrate at 160 °C, enables roughness values (R_a) as low as ~0.6 nm, which is in the same order of magnitude as the pristine substrate (~0.1 nm). The synthesis strategy is guided by *ab initio* calculations, indicating very strong interfacial bonding of the Cu nucleation layer to an α-Al_2O_3 substrate (work of separation 5.48 J/m^2)—which can be understood based on the high Cu coordination at the interface—and between *fcc* FeMn and Cu (3.45 J/m^2). Accompanied by small lattice misfits between these structures, the strong interfacial bonding is proposed to enable the local epitaxial growth of a smooth *fcc* FeMn thin film. Based on the here introduced synthesis strategy, the implementation of *fcc* FeMn based thin film model systems for materials with interface dominated properties such as FeMn steels containing κ-carbide precipitates or secondary phases appears meaningful.

Keywords: interphase; FeMn; density-functional theory

1. Introduction

The material combination FeMn in its face-centred cubic (*fcc*) phase plays an important role in structural materials such as high manganese steels, which exhibit excellent mechanical properties [1–4]. FeMn thin films can serve as model systems to gain fundamental understanding required for the design of those steels: Single phase model systems—deposited with the highly efficient combinatorial thin film composition-spread method and combined with *ab initio* calculations—were shown to be a powerful tool to rationalize composition-structure-property relationships [5]; Multiphase model systems with multilayers separated by flat (2D) interfaces are expected to render fundamental and systematic insights required for the design of materials with interface dominated properties, for example, FeMn steels containing κ-carbide precipitates [6,7]. Furthermore, FeMn thin films offer antiferromagnetic properties and are used in magnetic thin film devices, for example, combined with at least one ferromagnetic layer in a multilayer thin film exhibiting the so-called exchange bias (EB) effect [8–13].

The multi-layered FeMn based thin film model systems as well as magnetic devices require sharp interfaces for a meaningful determination and understanding of interface dominated properties like coherency effects [14] or EB [15]. The sputter deposition of FeMn based model systems on rigid oxide substrates requires relatively high deposition temperatures of up to 450 °C for the *fcc* phase formation, which is reported to be accompanied by high thin film roughness of up to 170 nm for film thicknesses in the order of 2 μm [16–19]. Due to the high surface roughness, the implementation of *fcc* FeMn thin films in layered thin film model systems did not yet appear meaningful. On the other hand, (ultra) thin *fcc* FeMn layers in magnetic devices with high quality interfaces are accomplished by stabilizing the *fcc* lattice by epitaxial thin film growth on metallic nucleation layers at low deposition temperatures [20–24] but layer thicknesses are restricted to several nm [22]. The template effect necessitates a low lattice mismatch between FeMn and the nucleation layer and thus a similar arrangement of the atoms at the interface [20,25]. The lattice parameter of *fcc* Cu (a_{Cu} = 3.615 Å [26]) is very similar to the lattice parameter of *fcc* FeMn (a_{FeMn} ≈ 3.59–3.74 Å [27]), whereby the latter is determined by the Fe to Mn ratio, giving rise to a low lattice mismatch <4% depending on the Mn content in a cube-on-cube oriented Cu/FeMn interface arrangement. If Cu were used as (bulk) substrate material, its mechanical properties (plastic deformation) would preclude analysis techniques such as nanoindentation. However, nanoindentation is possible if a thin Cu nucleation layer on a rigid (oxide) substrate material is employed. Schmidt et al. used MgO (a_{MgO} = 4.210 Å [28]) substrates for the molecular beam epitaxy (MBE) synthesis of the EB FeMn/Co. thin-films due to its cubic structure and lattice mismatch of about 16% for cube-on-cube orientation on FeMn [15]. Sapphire (α-Al$_2$O$_3$) substrates ($a_{\alpha-Al_2O_3}$ = 4.759 Å [29]) possess a hexagonal structure but exhibit a lattice mismatch of about 8% in (0001) stacking with Cu(111). Cu was shown to grow epitaxially on both oxide substrates: Cu(100) on MgO(100) [30] as well as Cu(111) on MgO(111) [30] or α-Al$_2$O$_3$ (0001) [31]. Moreover, Sigumonrong et al. have suggested that a small lattice misfit accompanied by strong interfacial bonding—characterized by a large work of separation—enables local epitaxial thin film growth and this notion was verified experimentally by the local epitaxial growth of 1.5 μm thick sputter deposited V$_2$AlC thin films on α-Al$_2$O$_3$(11$\bar{2}$0) [32]. Further examples for local epitaxial growth are Ti$_{1-x}$Al$_x$N [33] and Ti$_{1-x}$Al$_x$N/Ti$_{1-y}$Al$_y$N multilayer [34] thin films on ferritic steel substrates or α-(Cr,Al)$_2$O$_3$ thin films on α-Al$_2$O$_3$ substrates [35].

Thus, the goal of our paper is to develop a low temperature synthesis strategy based on local epitaxial growth on a Cu nucleation layer to enable the growth of smooth *fcc* FeMn films with surface roughness values in the order of magnitude of the pristine substrate. This development is guided by *ab initio* calculations identifying a layer system comprising a rigid substrate material, Cu nucleation layer and *fcc* FeMn thin film that is characterized by strong interfacial bonding across both interfaces. The work of separation for each interface in the layer systems MgO(001)/Cu(001)/Fe$_{0.6}$Mn$_{0.4}$(001) and α-Al$_2$O$_3$(0001)/Cu(111)/Fe$_{0.6}$Mn$_{0.4}$(111) is determined as a measure for the interfacial strength. The layer system α-Al$_2$O$_3$(0001)/Cu(111) will be demonstrated to exhibit the highest interfacial strength across the substrate/nucleation layer interface. Furthermore, the low temperature growth of a combinatorially deposited 100 nm thick FeMn layer onto the Cu nucleation layer is employed. This thin film exhibits a continuous Fe-Mn concentration gradient parallel to the substrate surface and thus allows to determine the Mn concentration range for the formation of *fcc* FeMn on the Cu nucleation layer.

2. Materials and Methods

To determine the work of separation between different layers of the systems MgO/Cu/FeMn and α-Al$_2$O$_3$/Cu/FeMn, density functional theory [36] was used. Vienna *ab initio* simulation package (VASP 5.4.1, Computational Materials Physics, University of Vienna, Vienna, Austria) and projector augmented wave potentials [37–39] were employed. The augmented wave potential parametrization was performed within the generalized-gradient approximation by Perdew, Burke and Ernzerhof [40]. The Blöchl approach [41] was applied for the total energy. The integration in the reciprocal space was carried out on 9 × 9 × 1 k-point Monkhorst-Pack mesh [42]. Fe$_{0.6}$Mn$_{0.4}$ was randomized using special quasirandom structures (SQS) [43], whereby locally self-consistent Green's function (LSGF) package [44,45] was employed. In particular, the Warren-Cowley short-range order parameter [46] within seven coordination shells was applied to treat randomness in these SQS configurations. The cells obtained from the LSGF code were used as input in the VASP code. $Fe_{0.6}Mn_{0.4}$ was spin polarized (antiferromagnetic ordering). Full structural optimization was made for each configuration employing the convergence criterion for the total energy of 0.01 meV and a 400 eV cut-off. Two interface configurations were studied: $MgO(001)/Cu(001)/Fe_{0.6}Mn_{0.4}(001)$ and α-$Al_2O_3(0001)/Cu(111)/Fe_{0.6}Mn_{0.4}(111)$. The interfaces were characterized by the work of separation [32,47], calculated from the total energy change per unit area upon separation of a slab from the remaining layered structure, for example, $Fe_{0.6}Mn_{0.4}(001)$ separation from $MgO(001)/Cu(001)$. $MgO(001)/Cu(001)/Fe_{0.6}Mn_{0.4}(001)$ was constructed by 6 layers of $MgO(001)$, 6 layers of $Cu(001)$ and 6 layers of $Fe_{0.6}Mn_{0.4}(001)$, accounting for 48 atoms. The layer thickness convergence was performed by considering 6 additional layers of $Cu(001)$. The work of separation differed by only 0.02 J/m^2 between 6 and 12 layer Cu configurations so that 6 layers were henceforth considered. The stacking sequence for $MgO(001)/Cu(001)$ was adopted from literature [48]. α-$Al_2O_3(0001)/Cu(111)/Fe_{0.6}Mn_{0.4}(111)$ was span by 7 layers of O terminated α-$Al_2O_3(0001)$, 6 layers of $Cu(111)$ and 6 layers of $Fe_{0.6}Mn_{0.4}(111)$. $Cu(111)$ and $Fe_{0.6}Mn_{0.4}(111)$ were constructed by A-B-C-A-B-C(111) stacking of the corresponding cubic lattice. The stacking of α-$Al_2O_3(0001)/Cu(111)$ was implemented based on a previous work by Hashibon et al. [49]. The number of layers implemented for the oxide substrates is assumed to be sufficient based on *ab initio* studies on the surface energy of the binary oxides by Lazar et al. [50] and Sigumonrong et al. [51]. All interfaces were constrained to the corresponding 0 K lattice parameters of the substrates (MgO and α-Al_2O_3) obtained herein.

The combinatorial Fe-Mn thin films on Cu nucleation layers were deposited in a laboratory-scale sputtering system by direct current magnetron sputtering. The base pressure was below 2.2×10^{-5} Pa and the Ar deposition pressure was 0.75 Pa. A Cu target (power density 0.86 W/cm^2) was facing the substrate for the deposition of the nucleation layers. A Fe target (power density 1.23 W/cm^2) and a Mn target (power density 0.36 W/cm^2) with inclination angles of $45°$ with respect to the substrate normal were used in order to achieve the desired Fe-Mn gradient. Fast acting target shutters (actuation time of 200 ms) ensured the deposition of defined Cu nucleation layers and the FeMn films directly followed by each other. Polished single-crystalline α-$Al_2O_3(0001)$ substrates in a target-to-substrate distance of 10 cm were kept at floating potential and heated to 160 °C during deposition.

The chemical composition of the combinatorial FeMn thin film was determined by energy dispersive X-ray analysis (EDX) in a JEOL JSM-6480 scanning electron microscope (JEOL Ltd., Tokyo, Japan) equipped with an EDAX Genesis 2000 EDX system (EDAX Inc., Mahwah, NJ, USA). Spatially resolved, high-throughput X-ray diffraction (XRD) measurements were used to evaluate the local crystal structure of the combinatorial Fe-Mn thin films using a Bruker AXS D8 Discover XRD (Bruker Corporation, Billerica, MA, USA) equipped with a General Area Detector Diffraction System (GADDS). The diffractometer was operated at a current of 40 mA and a voltage of 40 kV with Cu Kα radiation at a fixed incident angle ω of $8°$. ϕ-scans are performed with a frame width of $3°$ at $2\theta = 43.39°$, $\omega = 21.70°$ and $\chi = 20°$. The primary beam was collimated with a 0.5 mm pinhole. For Bragg-Brentano XRD measurements with a step size of $0.02°$, a Siemens D5000 (Siemens, Munich, Germany) diffractometer was employed and operated at a current of 40 mA and a voltage of 40 kV with Cu Kα radiation.

The surface roughness (R_a) of the FeMn thin film and a pristine α-$Al_2O_3(0001)$ substrate was determined using a Hysitron TI-950 TriboIndenter (Hysitron Inc., Eden Prairie, MN, USA) by scanning the film surface with a cube corner diamond tip and average the R_a values from ten areas of 10×10 μm^2 or six areas of 5×5 μm^2, respectively.

Site-specific lift-outs for microstructural characterization of the interface were carried out in a FEI Helios Nanolab 660 (FEI Company, Hillsboro, OR, USA) dual-beam focused ion beam (FIB) microscope. Lamellae were extracted in growth direction of the thin films and Ga ions were used at an acceleration voltage of 30 kV. First, a 1 μm thick Pt protection layer was applied, followed by trench milling, extraction of the lamella with a manipulation needle and application of the lamella on a Cu Omniprobe grid. Final lamella thicknesses were <100 nm. The lamellae were investigated using

EDX analysis in a JEOL JSM-2200FS field emission gun transmission electron microscope (TEM) (JEOL Ltd., Tokyo, Japan) in scanning TEM (STEM) mode, with a 30 mm diameter silicon drift detector from JEOL. In addition, selected area electron diffraction (SAED) investigations in the TEM were utilized to disclose the orientation dependency between the substrate and the Cu and FeMn layers. The results of the SAED were evaluated using the Java electron microscopy software (JEMS) (Version 3.8224U2012, JEMS-SAAS, Saas-Fee, Switzerland).

3. Results and Discussion

3.1. Ab Initio Interface Design

For the *ab initio* design of a suitable layer system, Figure 1 depicts the structural models as well as the corresponding electron density distributions, which are used to explore the electronic structure analysis. For MgO(001)/Cu(001)/Fe$_{0.6}$Mn$_{0.4}$(001), the calculated work of separation value is 0.85 J/m^2 for the substrate/Cu interface, which is consistent with literature [48] and 3.89 J/m^2 for the Cu/FeMn interface. For α-Al$_2$O$_3$(0001)/Cu(111)/Fe$_{0.6}$Mn$_{0.4}$(111), the work of separation value is 5.48 J/m^2 for the substrate/Cu interface, which is in line with previous studies [49] and 3.45 J/m^2 for the Cu/FeMn interface. Hence, striking differences occur for the considered substrate/Cu interfaces. MgO(001)/Cu(001) reveals a rather low work of separation, which is still by one order of magnitude higher compared to a weak interface, for example, ideal polypropylene/Ti$_{0.5}$Al$_{0.5}$N(002) [52] but smaller than medium strong interfaces like V$_2$AlC/α-Al$_2$O$_3$ [32] or Pt/NbO$_2$ [53]. The α-Al$_2$O$_3$(0001)/Cu(111) interface exhibits a more than six times higher work of separation which is in the range of strong interfaces such as Nb/α-Al$_2$O$_3$ [54], TiO$_2$/α-Al$_2$O$_3$ [55], or Al/diamond [56]. On the other hand, the work of separation for the Cu/Fe$_{0.6}$Mn$_{0.4}$ interface in both layer systems is comparable and differs by 0.44 J/m^2. The values are in the range of other medium strong interfaces such as Ir/Ir$_3$Zr [57].

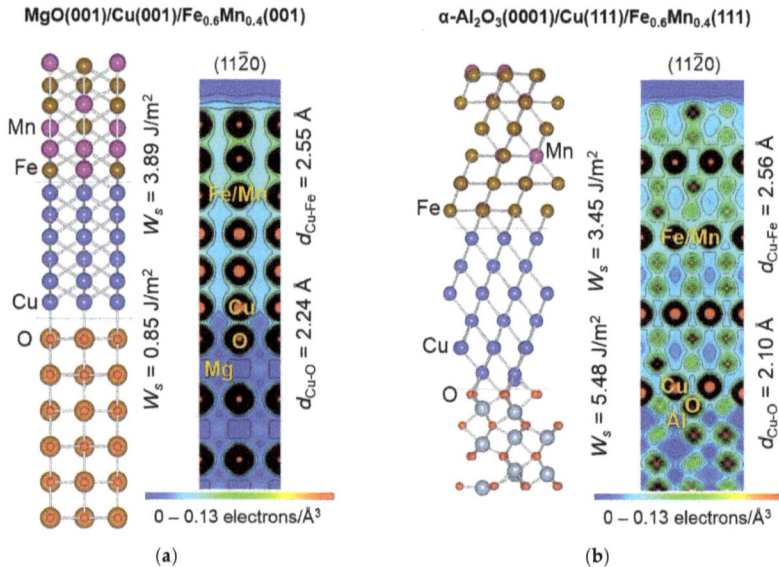

Figure 1. Structural models of (a) MgO(001)/Cu(001)/Fe$_{0.6}$Mn$_{0.4}$(001) and (b) α-Al$_2$O$_3$(0001)/Cu(111)/Fe$_{0.6}$Mn$_{0.4}$(111), each with corresponding electron density distribution in a (11$\bar{2}$0) plane of the substrate. The work of separation (W_S) and bond length (d) at the interfaces are also provided.

The distinct differences in the calculated work of separation for the substrate/Cu interface can be understood based on the Cu coordination at the interface, which is 1 between Cu and O for MgO(001)/Cu(001) but 3 for α-Al$_2$O$_3$(0001)/Cu(111) with Cu being bonded to O via (short and strong) ionic-covalent bonds. The origin of the difference in interfacial bonding can then be explained by the different electronic structures of MgO and α-Al$_2$O$_3$: The Mg-O bonds are predominantly ionic in character with significant charge transfer from Mg to O, while the Al-O bonds are characterized by (ionic) charge transfer as well as some localization (hybridization; consistent with literature on isostructural oxides [58]). For the O terminated α-Al$_2$O$_3$, the bond distance between Cu and O is 2.10 Å and the shared charge between Cu and O provides only Cu-O bonds in the interface. Compared to that, MgO is terminated by both O and Mg so that the weaker Cu-O interaction is characterized by less near neighbours and hence the lower coordination and a bond distance Cu-O of 2.25 Å.

The comparable work of separation values for the Cu/FeMn interfaces in both systems can be explained by the rather similar bonding conditions. The bond distance between Cu and Fe is 2.55 Å for Cu(001)/Fe$_{0.6}$Mn$_{0.4}$(001) compared to 2.56 Å for the Cu(111)/Fe$_{0.6}$Mn$_{0.4}$(111) interface exhibiting a 13% lower work of separation for the considered Fe$_{0.6}$Mn$_{0.4}$ lattice with its Mn content dependent lattice constant. As the nature of both the Cu as well as the Fe-Mn bonds are mainly metallic offering itinerant change with some localization between the atoms and there are no coordination modulations across the interface, hence, the bonding conditions at the Cu/FeMn interface are maintained throughout both layers.

Hence, the interfacial bonding between rigid substrate and Cu nucleation layer is theoretically shown to be extraordinary high for α-Al$_2$O$_3$(0001)/Cu(111), which may facilitate the epitaxial growth of Cu(111) on α-Al$_2$O$_3$(0001). The subsequent Cu(111)/Fe$_{0.6}$Mn$_{0.4}$(111) interface exhibits high interfacial strength as well and may thus allow for the continuation of the epitaxial growth throughout the interface between nucleation layer and FeMn thin film. Consequently, the layer system α-Al$_2$O$_3$(0001)/Cu(111)/Fe$_{0.6}$Mn$_{0.4}$(111) is selected to be appraised experimentally.

3.2. Experimental Validation

A 10 nm layer of Cu subsequently followed by a 100 nm combinatorial FeMn layer was synthesized on an α-Al$_2$O$_3$(0001) substrate and evaluated by EDX and spatially resolved XRD to identify the Mn content range, in which the desired *fcc* FeMn forms. Figure 2a shows the XRD scans (2θ-scans at fixed incidence angle ω) for Mn contents between 16 and 49 at. %. For Mn contents above 39 at. % the formation of phase pure *fcc* FeMn is observed. Lower Mn contents appear to facilitate the formation of an additional *bcc* phase and are thus excluded from further analysis. Figure 2b shows the Bragg-Brentano XRD scan of the *fcc* region of the α-Al$_2$O$_3$/Cu/FeMn sample, which is selected for TEM analysis (TEM-EDX: ~42 at. % Mn). For comparison, the XRD scan of a sole 10 nm Cu thin film (α-Al$_2$O$_3$/Cu)—deposited and measured under the same conditions—is included. Both Bragg-Brentano scans solely exhibit the (111) peak of the *fcc* structure. The intensity difference is caused by the difference in thin film thickness (110 nm for α-Al$_2$O$_3$/Cu/FeMn and 10 nm for α-Al$_2$O$_3$/Cu) and thus measured volume. Hence, the *fcc* FeMn layer as well as the Cu layer are highly (111) oriented and exhibit a very similar d_{111}-value of about 2.08 Å. The presence of a (111) fibre texture was excluded by ϕ-scans ($2\theta = 43.39°$, $\omega = 0°$, $\chi = 20°$) in the *fcc* region at ~42 at. % Mn. Figure S1 (Supplementary Material) shows the maximum intensities plotted in dependence of ϕ, exhibiting maximum intensities every 60° ϕ but minimum intensities in-between and thus refuting a fibre texture. Hence, XRD analysis indicates local epitaxial growth, which will be confirmed by high resolution techniques.

The surface roughness of the FeMn thin film is determined to be as low as $R_a = 0.6$ nm with a calculated standard deviation of 4% (based on ten individual measurements of 10×10 µm^2; a representative surface area scan can be found in the Supplementary Material, Figure S2). Hence, the surface roughness is on the same order of magnitude as the one measured for the pristine α-Al$_2$O$_3$ substrate ($R_a = 0.1$ nm). Thus, a significant reduction of the surface roughness compared to the high temperature synthesis strategy resulting in surface roughness values of R_a = 58 nm [19] and larger [16,17] (film thickness ~2 μm) could be obtained.

The STEM high angle annular dark field (HAADF) image of the layer system α-Al$_2$O$_3$/Cu/Fe$_{0.58}$Mn$_{0.42}$ with the corresponding EDX line scan shown in Figure 3 confirms the chemical stacking of the layer system: The α-Al$_2$O$_3$ substrate is characterized by high Al K and O K intensities, while the nucleation layer shows a peak in Cu K intensity for a length of about 10 nm. The Cu layer is directly followed by the ~100 nm thick FeMn layer with correspondingly high Fe K and Mn K intensities. The uppermost region of the FeMn layer already shows significant intensities of Pt M and Ga K, which is an effect of the FIB preparation necessitating the deposition of a Pt protection layer in order to reduce Ga implantation. Spot EDX measurements of the Fe and Mn content in the thin film layer result in 58 + 1 at. % and 42 ± 1 at. % Mn, respectively, which is close to the calculated Fe$_{0.6}$Mn$_{0.4}$ composition.

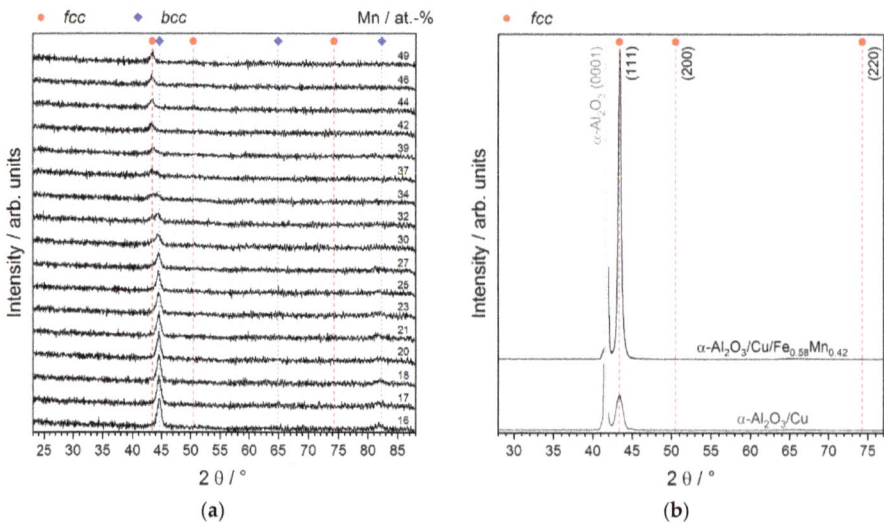

Figure 2. (a) X-ray diffraction (XRD) scans (2θ-scans at fixed incidence angle ω) of α-Al$_2$O$_3$/Cu/FeMn in dependence of the Mn content of the FeMn layer; (b) XRD scan (Bragg-Brentano) of α-Al$_2$O$_3$/Cu/Fe$_{0.58}$Mn$_{0.42}$, which is chosen for transmission electron microscopy (TEM) analysis (comparison: α-Al$_2$O$_3$/Cu without FeMn layer).

The SAED of the α-Al$_2$O$_3$/Cu/Fe$_{0.58}$Mn$_{0.42}$ layer system confirms that the α-Al$_2$O$_3$(0001) substrate is a chemically ordered structure and thus provides superstructure diffraction spots, as shown in the indexed diffraction pattern in Figure 4a. Figure 4b shows the diffraction pattern and an evaluation of the Cu/Fe$_{0.58}$Mn$_{0.42}$ layer. The Cu and the FeMn layer are not distinguishable and offer the same diffraction spots, confirming that they exhibit the same *fcc* structure as well as non-distinguishable lattice parameters. This was already indicated by the (macroscopic) XRD analysis discussed above, see Figure 2b. The *fcc* Fe$_{0.58}$Mn$_{0.42}$ is a chemically disordered solid solution, which is consistent with literature [25]. The orientation relationship between the α-Al$_2$O$_3$ substrate and the *fcc* Cu/Fe$_{0.58}$Mn$_{0.42}$ suggests (local) epitaxial growth and is determined as follows:

- α-Al$_2$O$_3$(0006) || Cu/Fe$_{0.58}$Mn$_{0.42}$($\bar{1}\bar{1}1$) [or ($11\bar{1}$)] and
- α-Al$_2$O$_3$($0\bar{3}30$) || Cu/Fe$_{0.58}$Mn$_{0.42}$($2\bar{2}0$) [or ($\bar{2}20$)]

The possibilities in square brackets are due to the rotational symmetry for a 180° rotation of the diffraction pattern. Furthermore, the inverted diffraction pattern of the layer system in Figure 4c reveals that the (small, focused) diffraction spots of the α-Al_2O_3 substrate are not positioned at the exact same positions as the (broad, unfocused) spots of the *fcc* Cu/$Fe_{0.58}Mn_{0.42}$ layer, indicating on the one hand that the (local) epitaxial growth is realized but also that the slightly different lattice plane spacing of α-Al_2O_3 and *fcc* Cu/$Fe_{0.58}Mn_{0.42}$ result in strain relaxation in the probed volume. Hence, we have demonstrated that *fcc* $Fe_{0.58}Mn_{0.42}(111)$ can be grown epitaxially on *fcc* Cu(111) which in turn grows epitaxially on α-$Al_2O_3(0001)$.

The low surface roughness of the 100 nm thick FeMn thin film, achieved by the local epitaxial growth on the Cu nucleation layer, allows for the formation of a flat (2D) interface between the *fcc* FeMn if another layer is deposited on top. This is a necessary step towards realizing the deposition of a bi- or multi-layered (multiphase) thin film model systems to allow for the systematic evaluation of interface dominated properties. These model systems are relevant to explore for example FeMn steels containing precipitates or secondary phases. The single phase model systems can benefit from the low surface roughness in terms of for example, a more accurate determination of mechanical properties by nanoindentation or other roughness-sensitive analysis techniques.

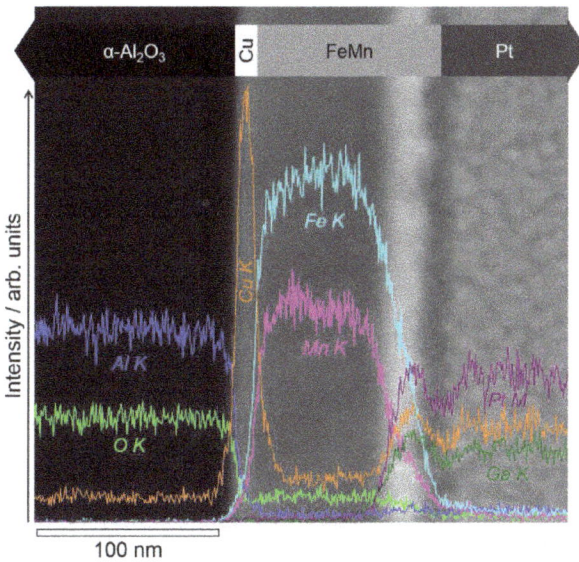

Figure 3. Scanning transmission electron microscopy (STEM) high angle annular dark field (HAADF) image of the layer system α-Al_2O_3/Cu/$Fe_{0.58}Mn_{0.42}$ with corresponding energy dispersive X-ray (EDX) line scan including the Al K, O K, Cu K, Fe K, Mn K, Pt M and Ga K intensities. The intensity signal is not corrected for peak overlapping.

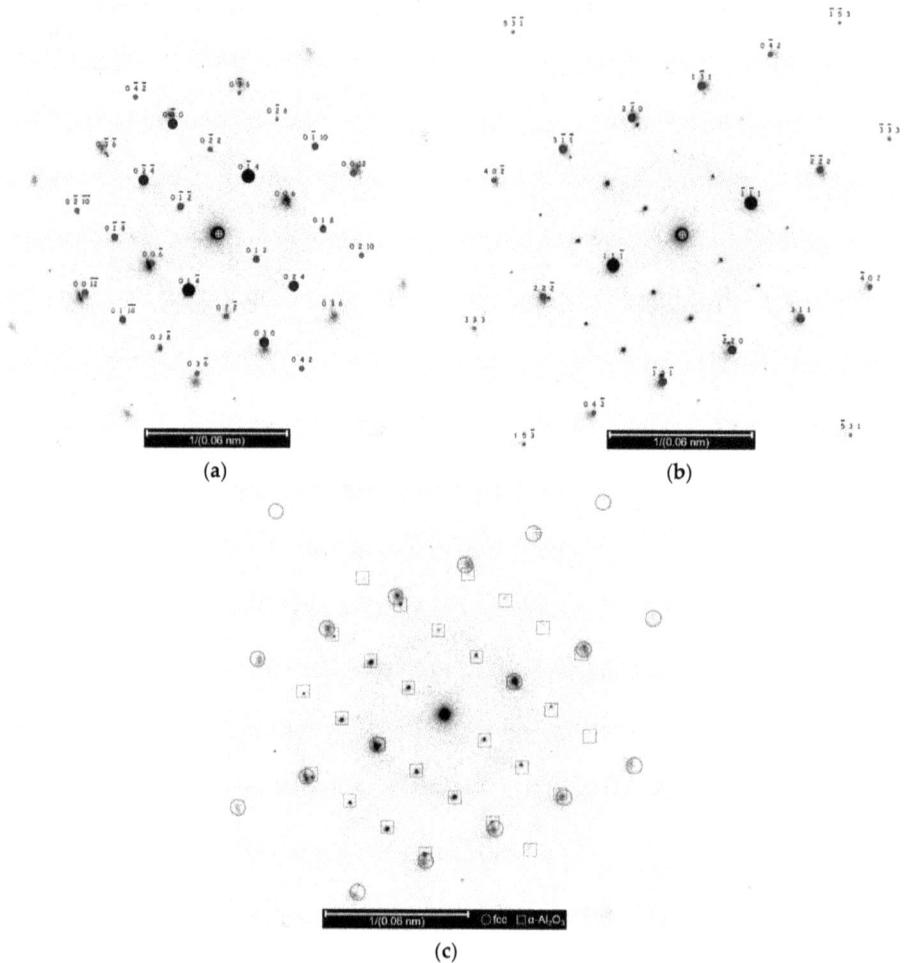

Figure 4. Indexed diffraction pattern of (**a**) α-Al$_2$O$_3$ substrate (zone axis $[2\bar{1}\bar{1}0]$); (**b**) *fcc* Cu/Fe$_{0.58}$Mn$_{0.42}$ layer (zone axis [112]); (**c**) Diffraction pattern of the layer system α-Al$_2$O$_3$/Cu/Fe$_{0.58}$Mn$_{0.42}$ (*fcc* and α-Al$_2$O$_3$ diffraction spots are labelled with red circles and blue squares, respectively).

4. Conclusions

In conclusion, a low temperature deposition strategy for the growth of smooth *fcc* FeMn thin films was developed. *Ab initio* calculations suggest that the layer structure α-Al$_2$O$_3$(0001)/Cu(111) exhibits higher interfacial strength than MgO(001)/Cu(001) and is thus chosen as substrate/nucleation layer combination to enable (local) epitaxial growth of *fcc* Cu(111) on α-Al$_2$O$_3$(0001) as well as (local) epitaxial growth of *fcc* FeMn(111) on *fcc* Cu(111). The work of separation between Cu and Fe$_{0.6}$Mn$_{0.4}$ does not differ significantly between the two configurations studied, and, as such, is not decisive regarding the selection of the substrate system. However, the α-Al$_2$O$_3$(0001)/Cu(111) interface exhibits a more than six times higher work of separation (5.48 J/m^2) than the MgO(001)/Cu(001) interface (0.85 J/m^2) and hence strong interfacial bonding. This distinct difference in the calculated work of separation can be understood based on the Cu coordination at the interface, which is 1 between Cu and O for MgO(001)/Cu(001) but 3 for α-Al$_2$O$_3$(0001)/Cu(111).

The low temperature deposition strategy for the growth of smooth *fcc* FeMn thin films was critically appraised by synthesizing a 100 nm thick (combinatorial) FeMn thin film on a 10 nm Cu nucleation layer which was in turn deposited on an α-Al$_2$O$_3$(0001) substrate (α-Al$_2$O$_3$/Cu/FeMn). XRD confirms the phase formation of *fcc* FeMn for Mn contents above 39 at. %. The (local) epitaxial growth of *fcc* Fe$_{0.58}$Mn$_{0.42}$(111) on *fcc* Cu(111) on α-Al$_2$O$_3$(0001) is verified by TEM/SAED with the following orientation relationship: α-Al$_2$O$_3$(0006) || Cu/Fe$_{0.58}$Mn$_{0.42}$($\overline{1}\overline{1}1$) and α-Al$_2O_3$($0\overline{3}30$) || Cu/Fe$_{0.58}Mn_{0.42}$($2\overline{2}0$). The thin film exhibits the desired low surface roughness with R$_a$ values as low as 0.6 nm, which is on the same order of magnitude as the α-Al$_2$O$_3$ substrate (R$_a$ ~0.1 nm) and at least two orders of magnitude lower than the *fcc* FeMn based thin films deposited at high temperatures (R$_a \geq$ 58 nm). Thus, the introduced synthesis strategy opens the possibility of the successful implementation of thin film model systems for materials with interface dominated properties such as FeMn based steels containing κ-carbide precipitates.

Supplementary Materials: The following are available online at http://www.mdpi.com/2075-4701/8/6/384/s1.

Author Contributions: F.H., D.M. and J.M.S. conceived and designed the experiments and DFT calculations; D.M. designed and performed the DFT calculations; F.H., P.J.P. and A.L.R. performed the sputtering experiments as well as XRD, EDX and surface roughness analysis; M.H. performed the TEM sample preparation and B.V. conceived and performed the TEM analysis; D.M. and J.M.S. initiated and supervised the project; all authors contributed to the interpretation of the data and the writing of the paper.

Funding: This research was funded by the Deutsche Forschungsgemeinschaft (DFG) within the Collaborative Research Center (SFB) 761 "Steel *ab initio*".

Acknowledgments: Simulations were performed with computing resources granted by JARA-HPC from RWTH Aachen University under project JARA0152.

Conflicts of Interest: The authors declare no conflict of interest.

References

1. Frommeyer, G.; Brux, U.; Neumann, P. Supra-ductile and high-strength manganese-TRIP/TWIP steels for high energy absorption purposes. *ISIJ Int.* **2003**, *43*, 438–446. [CrossRef]
2. Peng, C.T.; Callaghan, M.D.; Li, H.J.; Yan, K.; Liss, K.D.; Ngo, T.D.; Mendis, P.A.; Choi, C.H. On the compression behavior of an austenitic Fe-18Mn-0.6C-1.5Al twinning-induced plasticity steel. *Steel Res. Int.* **2013**, *84*, 1281–1287. [CrossRef]
3. Gutierrez-Urrutia, I.; Raabe, D. Dislocation and twin substructure evolution during strain hardening of an Fe-22 wt. % Mn-0.6 wt. % C TWIP steel observed by electron channeling contrast imaging. *Acta Mater.* **2011**, *59*, 6449–6462. [CrossRef]
4. Gebhardt, T.; Music, D.; Hallstedt, B.; Ekholm, M.; Abrikosov, I.A.; Vitos, L.; Schneider, J.M. Ab initio lattice stability of *fcc* and hcp Fe-Mn random alloys. *J. Phys. Condes. Matter* **2010**, *22*, 295402. [CrossRef] [PubMed]
5. Gebhardt, T.; Music, D.; Takahashi, T.; Schneider, J.M. Combinatorial thin film materials science: From alloy discovery and optimization to alloy design. *Thin Solid Films* **2012**, *520*, 5491–5499. [CrossRef]
6. Gutierrez-Urrutia, I.; Raabe, D. Influence of Al content and precipitation state on the mechanical behavior of austenitic high-Mn low-density steels. *Scr. Mater.* **2013**, *68*, 343–347. [CrossRef]
7. Timmerscheidt, T.; Dey, P.; Bogdanovski, D.; von Appen, J.; Hickel, T.; Neugebauer, J.; Dronskowski, R. The role of κ-carbides as hydrogen traps in high-Mn steels. *Metals* **2017**, *7*, 264. [CrossRef]
8. Kuch, W.; Chelaru, L.I.; Offi, F.; Wang, J.; Kotsugi, M.; Kirschner, J. Tuning the magnetic coupling across ultrathin antiferromagnetic films by controlling atomic-scale roughness. *Nat. Mater.* **2006**, *5*, 128–133. [CrossRef] [PubMed]
9. Lenssen, K.M.H.; van Kesteren, H.W.; Rijks, T.; Kools, J.C.S.; de Nooijer, M.C.; Coehoorn, R.; Folkerts, W. Giant magnetoresistance and its application in recording heads. *Sens. Actuator A Phys.* **1997**, *60*, 90–97. [CrossRef]
10. Savin, P.; Guzman, J.; Lepalovskij, V.; Svalov, A.; Kurlyandskaya, G.; Asenjo, A.; Vas'kovskiy, V.; Vazquez, M. Exchange bias in sputtered FeNi/FeMn systems: Effect of short low-temperature heat treatments. *J. Magn. Magn. Mater.* **2016**, *402*, 49–54. [CrossRef]

11. Svalov, A.; Savin, P.; Lepalovskij, V.; Larranaga, A.; Vas'kovskiy, V.; Garcia-Arribas, A.; Kurlyandskaya, G. Tailoring the exchange bias in FeNi/FeMn bilayers by heat treatment and FeMn surface oxidation. *IEEE Trans. Magn.* **2014**, *50*, 1–4. [CrossRef]

12. Stocks, G.M.; Shelton, W.A.; Schulthess, T.C.; Ujfalussy, B.; Butler, W.H.; Canning, A. On the magnetic structure of gamma-FeMn alloys. *J. Appl. Phys.* **2002**, *91*, 7355–7357. [CrossRef]

13. Binasch, G.; Grünberg, P.; Saurenbach, F.; Zinn, W. Enhanced magnetoresistance in layered magnetic structures with antiferromagnetic interlayer exchange. *Phys. Rev. B* **1989**, *39*, 4828–4830. [CrossRef]

14. Sawada, H.; Taniguchi, S.; Kawakami, K.; Ozaki, T. Transition of the interface between iron and carbide precipitate from coherent to semi-coherent. *Metals* **2017**, *7*, 277. [CrossRef]

15. Schmidt, M.; Grafe, J.; Audehm, P.; Phillipp, F.; Schutz, G.; Goering, E. Preparation and characterisation of epitaxial Pt/Cu/FeMn/Co thin films on (100)-oriented MgO single crystals. *Phys. Status Solidi A* **2015**, *212*, 2114–2123. [CrossRef]

16. Reeh, S.; Music, D.; Gebhardt, T.; Kasprzak, M.; Japel, T.; Zaefferer, S.; Raabe, D.; Richter, S.; Schwedt, A.; Mayer, J.; et al. Elastic properties of face-centred cubic Fe-Mn-C studied by nanoindentation and ab initio calculations. *Acta Mater.* **2012**, *60*, 6025–6032. [CrossRef]

17. Reeh, S.; Kasprzak, M.; Klusmann, C.D.; Stalf, F.; Music, D.; Ekholm, M.; Abrikosov, I.A.; Schneider, J.M. Elastic properties of *fcc* Fe-Mn-X (X = Cr, Co, Ni, Cu) alloys studied by the combinatorial thin film approach and ab initio calculations. *J. Phys. Condens. Matter* **2013**, *25*, 245401. [CrossRef] [PubMed]

18. Gebhardt, T.; Music, D.; Ekholm, M.; Abrikosov, I.A.; von Appen, J.; Dronskowski, R.; Wagner, D.; Mayer, J.; Schneider, J.M. Influence of chemical composition and magnetic effects on the elastic properties of *fcc* Fe-Mn alloys. *Acta Mater.* **2011**, *59*, 1493–1501. [CrossRef]

19. Gebhardt, T.; Music, D.; Kossmann, D.; Ekholm, M.; Abrikosov, I.A.; Vitos, L.; Schneider, J.M. Elastic properties of *fcc* Fe-Mn-X (X = Al, Si) alloys studied by theory and experiment. *Acta Mater.* **2011**, *59*, 3145–3155. [CrossRef]

20. Offi, F.; Kuch, W.; Kirschner, J. Structural and magnetic properties of Fe$_x$Mn$_{1-x}$ thin films on Cu(001) and on Co/Cu(001). *Phys. Rev. B* **2002**, *66*, 064419. [CrossRef]

21. Kuch, W.; Chelaru, L.I.; Kirschner, J. Surface morphology of antiferromagnetic Fe$_{50}$Mn$_{50}$ layers on Cu(001). *Surf. Sci.* **2004**, *566–568*, 221–225. [CrossRef]

22. Wuttig, M.; Feldmann, B.; Flores, T. The correlation between structure and magnetism for ultrathin metal films and surface alloys. *Surf. Sci.* **1995**, *331–333*, 659–672. [CrossRef]

23. Allegranza, O.; Chen, M.M. Effect of substrate and antiferromagnetic films thickness on exchange-bias field (invited). *J. Appl. Phys.* **1993**, *73*, 6218–6222. [CrossRef]

24. Sankaranarayanan, V.K.; Yoon, S.M.; Kim, D.Y.; Kim, C.O.; Kim, C.G. Exchange bias in NiFe/FeMn/NiFe trilayers. *J. Appl. Phys.* **2004**, *96*, 7428–7434. [CrossRef]

25. Ekholm, M.; Abrikosov, I.A. Structural and magnetic ground-state properties of gamma-FeMn alloys from ab initio calculations. *Phys. Rev. B* **2011**, *84*, 104423. [CrossRef]

26. *Cu Crystal Structure*; Datasheet from "Pauling file Multinaries Edition—2012" in Springermaterials; Springer: Berlin/Heidelberg, Germany; Material Phases Data System (MPDS): Vitznau, Switzerland; National Institute for Materials Science (NIMS): Tsukuba, Japan. Available online: http://materials.Springer.Com/isp/crystallographic/docs/sd_0250160 (accessed on 28 November.2017).

27. Predel, B. Fe-Mn (iron-manganese): Datasheet from landolt-börnstein—Group iv physical chemistry volume 5e: "Dy-Er—Fr-Mo". In *SpringerMaterials*; Springer: Berlin/Heidelberg, Germany, 1995.

28. *MgO Crystal Structure*; Datasheet from "Pauling File Multinaries Edition—2012" in Springermaterials; Springer: Berlin/Heidelberg, Germany; Material Phases Data System (MPDS): Vitznau, Switzerland; National Institute for Materials Science (NIMS): Tsukuba, Japan. Available online: http://materials.Springer.Com/isp/crystallographic/docs/sd_0305005 (accessed on 28 November 2017).

29. *α-Al$_2$O$_3$ (Al$_2$O$_3$ Cor) Crystal Structure*; Datasheet from "Pauling File Multinaries Edition—2012" in Springermaterials; Springer: Berlin/Heidelberg, Germany; Material Phases Data System (MPDS): Vitznau, Switzerland; National Institute for Materials Science (NIMS): Tsukuba, Japan. Available online: http://materials.Springer.Com/isp/crystallographic/docs/sd_0453386 (accessed on 28 November 2017).

30. He, J.-W.; Møller, P.J. Epitaxial and electronic structures of ultra-thin copper films on MgO crystal surfaces. *Surf. Sci.* **1986**, *178*, 934–942. [CrossRef]

31. Oh, S.H.; Scheu, C.; Wagner, T.; Tchernychova, E.; Ruhle, M. Epitaxy and bonding of Cu films on oxygen-terminated alpha-Al$_2$O$_3$(0001) surfaces. *Acta Mater.* **2006**, *54*, 2685–2696. [CrossRef]

32. Sigumonrong, D.P.; Zhang, J.; Zhou, Y.; Music, D.; Emmerlich, J.; Mayer, J.; Schneider, J.M. Interfacial structure of V$_2$AlC thin films deposited on (11-20)-sapphire. *Scr. Mater.* **2011**, *64*, 347–350. [CrossRef]

33. Schönjahn, C.; Donohue, L.A.; Lewis, D.B.; Münz, W.-D.; Twesten, R.D.; Petrov, I. Enhanced adhesion through local epitaxy of transition-metal nitride coatings on ferritic steel promoted by metal ion etching in a combined cathodic arc/unbalanced magnetron deposition system. *J. Vac. Sci. Technol. A* **2000**, *18*, 1718–1723. [CrossRef]

34. Petrov, I.; Losbichler, P.; Bergstrom, D.; Greene, J.E.; Münz, W.D.; Hurkmans, T.; Trinh, T. Ion-assisted growth of Ti$_{1-x}$Al$_x$N/Ti$_{1-y}$Nb$_y$N multilayers by combined cathodic-arc/magnetron-sputter deposition. *Thin Solid Films* **1997**, *302*, 179–192. [CrossRef]

35. Gao, Y.; Leiste, H.; Ulrich, S.; Stueber, M. Synthesis of local epitaxial α-(Cr$_{1-x}$Al$_x$)$_2$O$_3$ thin films (0.08 ≤ x ≤ 0.16) on α-Al$_2$O$_3$ substrates by r.f. Magnetron sputtering. *Thin Solid Films* **2017**, *644*, 129–137. [CrossRef]

36. Hohenberg, P.; Kohn, W. Inhomogeneous electron gas. *Phys. Rev.* **1964**, *136*, B864. [CrossRef]

37. Kresse, G.; Hafner, J. Ab initio molecular dynamics for open-shell transition metals. *Phys. Rev. B* **1993**, *48*, 13115. [CrossRef]

38. Kresse, G.; Hafner, J. Ab initio molecular-dynamics simulation of the liquid-metal-amorphous-semiconductor transition in germanium. *Phys. Rev. B* **1994**, *49*, 14251. [CrossRef]

39. Kresse, G.; Joubert, D. From ultrasoft pseudopotentials to the projector augmented wave method. *Phys. Rev. B* **1999**, *59*, 1758. [CrossRef]

40. Perdew, J.P.; Burke, K.; Ernzerhof, M. Generalized gradient approximation made simple. *Phys. Rev. Lett.* **1996**, *77*, 3865. [CrossRef] [PubMed]

41. Blöchl, P.E. Projector augmented-wave method. *Phys. Rev. B* **1994**, *50*, 17953. [CrossRef]

42. Monkhorst, H.J.; Pack, J.D. Special points for brillouin-zone integrations. *Phys. Rev. B* **1976**, *13*, 5188. [CrossRef]

43. Zunger, A.; Wei, S.-H.; Ferreira, L.G.; Bernard, J.E. Special quasirandom structures. *Phys. Rev. Lett.* **1990**, *65*, 353. [CrossRef] [PubMed]

44. Abrikosov, I.A.; Niklasson, A.M.N.; Simak, S.I.; Johansson, B.; Ruban, A.V.; Skriver, H.L. Order-N green's function technique for local environment effects in alloys. *Phys. Rev. Lett.* **1996**, *76*, 4203. [CrossRef] [PubMed]

45. Abrikosov, I.A.; Simak, S.I.; Johansson, B.; Ruban, A.V.; Skriver, H.L. Locally self-consistent green's function approach to the electronic structure problem. *Phys. Rev. B* **1997**, *56*, 9319. [CrossRef]

46. Cowley, J.M. X-ray measurement of order in single crystals of Cu$_3$Au. *J. Appl. Phys.* **1950**, *21*, 24–30. [CrossRef]

47. Lin, Z.; Bristowe, P.D. Microscopic characteristics of the Ag(111)/ZnO(0001) interface present in optical coatings. *Phys. Rev. B* **2007**, *75*, 205423. [CrossRef]

48. Matsunaka, D.; Shibutani, Y. Electronic states and adhesion properties at metal/MgO incoherent interfaces: First-principles calculations. *Phys. Rev. B* **2008**, *77*, 165435. [CrossRef]

49. Hashibon, A.; Elsässer, C.; Rühle, M. Structure at abrupt copper–alumina interfaces: An ab initio study. *Acta Mater.* **2005**, *53*, 5323–5332. [CrossRef]

50. Lazar, P.; Otyepka, M. Accurate surface energies from first principles. *Phys. Rev. B* **2015**, *91*, 115402. [CrossRef]

51. Sigumonrong, D.P.; Music, D.; Schneider, J.M. Efficient supercell design for surface and interface calculations of hexagonal phases: α -Al$_2$O$_3$ case study. *Comput. Mater. Sci.* **2011**, *50*, 1197–1201. [CrossRef]

52. Music, D.; Lange, D.; Raumann, L.; Baben, M.t.; von Fragstein, F.; Schneider, J.M. Polypropylene–MALN (M = Ti, Cr) interface interactions. *Surf. Sci.* **2012**, *606*, 986–989. [CrossRef]

53. Music, D.; Schmidt, P.; Saksena, A. Experimental and theoretical exploration of mechanical stability of Pt/NbO$_2$ interfaces for thermoelectric applications. *J. Phys. D Appl. Phys.* **2017**, *50*, 455502. [CrossRef]

54. Batirev, I.G.; Alavi, A.; Finnis, M.W.; Deutsch, T. First-principles calculations of the ideal cleavage energy of bulk niobium(111)/α-alumina(0001) interfaces. *Phys. Rev. Lett.* **1999**, *82*, 1510–1513. [CrossRef]

55. Popov, M.N.; Spitaler, J.; Mühlbacher, M.; Walter, C.; Keckes, J.; Mitterer, C.; Draxl, C. TiO$_2$(100)/Al$_2$O$_3$(0001) interface: A first-principles study supported by experiment. *Phys. Rev. B* **2012**, *86*, 205309. [CrossRef]

56. Qi, Y.; Hector, L.G. Hydrogen effect on adhesion and adhesive transfer at aluminum/diamond interfaces. *Phys. Rev. B* **2003**, *68*, 201403. [CrossRef]
57. Gong, H.R.; Liu, Y.; Tang, H.P.; Xiang, C.S. Bond strength and electronic structures of coherent Ir/Ir$_3$Zr interfaces. *Appl. Phys. Lett.* **2008**, *92*, 211914. [CrossRef]
58. Eyert, V.; Schwingenschlögl, U.; Eckern, U. Covalent bonding and hybridization effects in the corundum-type transition-metal oxides V$_2$O$_3$ and Ti$_2$O$_3$. *EPL (Europhys. Lett.)* **2005**, *70*, 782. [CrossRef]

![metals logo] *metals*

MDPI

Article

Multiscale Modelling of Hydrogen Transport and Segregation in Polycrystalline Steels

Claas Hüter [1,2,3,*], Pratheek Shanthraj [1], Eunan McEniry [1], Robert Spatschek [1,2,3], Tilmann Hickel [1], Ali Tehranchi [1], Xiaofei Guo [4] and Franz Roters [1]

[1] Max-Planck-Institut für Eisenforschung, Max-Planck-Straße 1, 40237 Düsseldorf, Germany; p.shanthraj@mpie.de (P.S.); e.mceniry@mpie.de (E.M.); r.spatschek@fz-juelich.de (R.S.); t.hickel@mpie.de (T.H.); tehranchi@mpie.de (A.T.); f.roters@mpie.de (F.R.)
[2] Institute for Energy and Climate Research, Forschungszentrum Jülich, 52425 Jülich, Germany
[3] Jülich-Aachen Research Alliance (JARA Energy), RWTH Aachen University, 52056 Aachen, Germany
[4] Steel Institute, RWTH Aachen University, 52072 Aachen, Germany; Xiaofei.Guo@iehk.rwth-aachen.de
* Correspondence: c.hueter@fz-juelich.de; Tel.: +49-2461-61-1569

Received: 30 April 2018; Accepted: 4 June 2018; Published: 7 June 2018

Abstract: A key issue in understanding and effectively managing hydrogen embrittlement in complex alloys is identifying and exploiting the critical role of the various defects involved. A chemo-mechanical model for hydrogen diffusion is developed taking into account stress gradients in the material, as well as microstructural trapping sites such as grain boundaries and dislocations. In particular, the energetic parameters used in this coupled approach are determined from ab initio calculations. Complementary experimental investigations that are presented show that a numerical approach capable of massive scale-bridging up to the macroscale is required. Due to the wide range of length scales accounted for, we apply homogenisation schemes for the hydrogen concentration to reach simulation dimensions comparable to metallurgical process scales. Via a representative volume element approach, an ab initio based scale bridging description of dislocation-induced hydrogen aggregation is easily accessible. When we extend the representative volume approach to also include an analytical approximation for the ab initio based description of grain boundaries, we find conceptual limitations that hinder a quantitative comparison to experimental data in the current stage. Based on this understanding, the development of improved strategies for further efficient scale bridging approaches is foreseen.

Keywords: hydrogen embrittlement; multi-scale; multiscale modelling; chemo-mechanics

---

1. Introduction

Hydrogen embrittlement (HE) can be defined as the structural degradation of materials resulting from exposure to hydrogen and often leading to abrupt and premature failure [1–7]. HE in complex engineering materials is increasingly commonplace in key application areas, such as hydrogen-based energy conversion cycles [8,9], high strength materials synthesis and coatings [10], marine and deep sea technology as well as structural components in the oil and gas industries [11]. While there is a critical need for such technologies to meet increasing worldwide energy demands, the damage risks associated with HE has become a substantial bottleneck for further development.

Despite considerable effort, the mechanistic causes for HE are not yet completely understood owing to its complex nature [12–14]. The competition of different defects for aggregating hydrogen, especially between dislocations and grain boundaries, poses a crucial as well as intricate subject of investigation in hydrogen embrittlement. Especially for dislocations and grain boundaries, the competitive picture is complemented by possibly cooperative damage facilitation, as hydrogen transport to grain boundaries via slip transfer. Currently, some viable mechanisms proposed in the literature are: (i) hydrogen enhanced de-cohesion (HEDE), (ii) hydrogen enhanced localized plasticity (HELP), (iii) hydride-induced embrittlement, and (iv) hydrogen induced super-abundant vacancy (HISAV) formation. In the HEDE mechanism, hydrogen diffusion to and its subsequent interaction with the strained atomic bonds at the crack tip, result in a lowering of the cohesive energy of the material [15], thus making it easier to form a Griffith crack [16–18]. The HELP mechanism was proposed to account for observations of localized plasticity at the crack tip in a range of metallic systems [2,19]. It is based on the influence of hydrogen in reducing the mobility of dislocations by screening their interaction stress fields [20,21]. The corresponding increase in plasticity is highly localized owing to the heterogeneous hydrogen distribution due to stress concentrations in the material, and thus culminates prematurely in ductile fracture.

These failure mechanisms and failure-inducing effects depend on localised chemistry, stress and deformation state, and the defects accessible to damage initiation. We therefore aim at the development of an efficient, massively scale-bridging approach that allows for catching process-relevant states of steel products, including massive deformations, high defect densities and varying chemical loadings. The approach of choice is a coupled crystal plasticity finite element (CPFEM) phase field model that is implemented in the Düsseldorf Advanced Material Simulation Kit (DAMASK) [22]. It operates on the macroscale and therefore employs representative volume element (RVE) descriptions that allow for including electronic and atomistic scale information about the system via efficient averaging in terms of composite models.

The manuscript is organised as follows: in Section 2, our recent experimental findings in ferritic steels, which exhibit grain boundary related hydrogen embrittlement, are presented. Hereby, we motivate the subsequent theoretical considerations. In Section 3, the continuum model is introduced, followed by a brief description of the implementation and the ab initio-based parametrisation. In Section 4, the analytic ab initio based approximations for the influence of dislocations on hydrogen aggregation are introduced and we compare the analytic composite model and the fully numerically resolved full-field simulation results. Based on the excellent agreement, we extend the composite model by an efficient, ab initio based description of grain boundaries. We recognise the conceptual difficulties to include grain boundaries in terms of a composite model in an RVE spirit. Finally, in Section 5, we summarise the insights gained and relate them to potential future approaches to the inherently massively scale bridging problem of hydrogen embrittlement.

2. Experimental Findings

Our theoretical considerations are complemented by experimental measurements of hydrogen-charged martensitic steel samples, focusing on fracture strength and hydrogen saturation. As we are interested in a qualitative estimate for the difference in hydrogen concentration in steel due to defects relative to a basic Sievert's law estimate, a cold rolled martensitic stainless steel X20Cr13 is chosen. Here, we expect a high dislocation density and a pronounced effect of defect-mediated hydrogen aggregation.

Figure 1 shows an example of hydrogen induced cracking in a cold rolled martensitic stainless steel X20Cr13 with the chemical composition 0.2% C and 13% Cr in weight percent. The material was pre-charged with hydrogen in 0.05 M H_2SO_4 and 1.4 g/L Thiourea at the overpotential of -800 mV$_{SCE}$ from 0–24 h to bring in different amounts of hydrogen into the material. Hydrogen charging was carried out with potentiostat High-Power 96 from Bank GmbH, Pohlheim, Germany . A Calomel electrode was used as a reference electrode and a platinum net was used as the counter electrode. Before charging, the specimen surfaces were ground sequentially from #320 to #800 SiC grit paper and finally polished with 6 μm diamond paste on canvas. After charging, the material was stored in liquid N_2 before hydrogen measurement by hot extraction with the hydrogen analysis equipment LECO RH402. Figure 1a shows the evolution of hydrogen contents according to the charging period. It reveals the material obtained as high as 27 ppm hydrogen after 24 h hydrogen charging. According to the Sievert's law, the solid solubility of hydrogen in pure iron under hydrogen pressure of 100 bar at room temperature accounts for 2×10^{-7}, which is extremely low compared to the amount of hydrogen measured in this experiment [23]. Therefore, it is assumed the oversaturated hydrogen due to the hydrogen charging is associated with hydrogen at dislocations and grain boundaries. Figure 1a also reveals the sharp reduction of fracture strength due to the charged hydrogen after a slow strain rate test at the strain rate of 10^{-6} s^{-1}, which is reduced from 1046 MPa to 432 MPa when the amount of hydrogen is raised from 1.5 ppm to 20 ppm. Slow strain rate tests were performed with a constant extension machine from Zwick, Ulm, Germany, with a maximum load of 30 kN. The tensile specimens are in dog-bone shape with the geometry A25. Figure 1b illustrates the fracture surface from the failed slow strain rate specimens. The as delivered X20Cr13 has the initial hydrogen content of 1.5 ppm, which exhibits fully ductile (D) fracture feature with very fine dimple sizes. After charging with 5 ppm hydrogen, the fracture surface changes to mixed transgranular (TG) and intergranular (IG) cleavage and a few ductile islands. The cracks are propagating through the prior austenite grain boundaries as well as the martensite lath. By further increasing the charged hydrogen, the TG and IG fractures become more prominent. In association with the sharp reduction in fracture strength, the high amount of charged hydrogen is assumed to be accumulating at the prior austenite grain boundaries and reduces the grain boundary cohesion force.

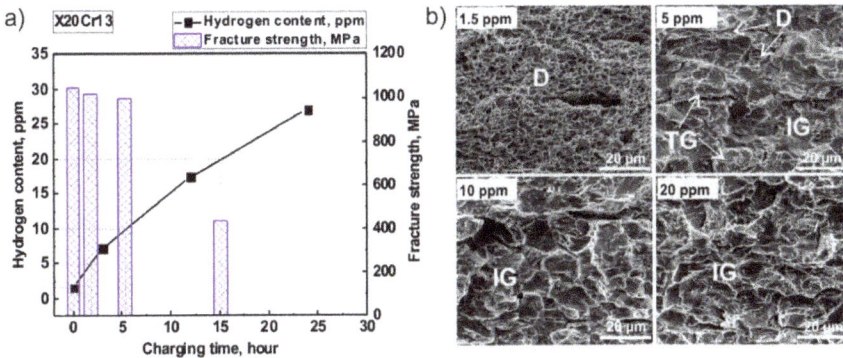

Figure 1. Hydrogen induced cracking in a cold rolled X20Cr13 martensitic stainless steel. (**a**) the evolution of hydrogen contents according to the hydrogen charging time and the fracture strength according to the hydrogen charging time determined by a slow strain rate test at the strain rate of 10^{-6} s^{-1}; (**b**) fracture surfaces from the failed slow strain rate specimens with different amounts of pre-charged hydrogen.

The complexity of the possible interpretation of the reported observations suggests a theoretical approach that reaches the experimental dimensions. On the other hand, it also needs to catch essential aspects of atomistic and electronic interaction of the hydrogen with the dominant species, here presumably dislocations and grain boundaries. In the following section, the ab initio informed crystal plasticity model is introduced, which we extend to include the effective description of these defect species in the context of hydrogen aggregation.

3. Model Formulation

The description of the entire model covers several aspects on multiple length scales. For an extended discussion of the model, see [24]. Following the continuum picture of the crystal plasticity model and hydrogen transport, the numerical implementation is briefly introduced. Based on the continuum picture, we explain which insights and parameters from the quantum mechanical level we use for the scale-bridging approach, and briefly discuss the ab initio model.

3.1. Continuum Model Formulation

Let $\mathcal{B}_0 \subset \mathbb{R}^3$ be a microstructural domain of interest, with boundary $\partial\mathcal{B}_0$. The deformation resulting from an applied loading defines a field, $\chi(\mathbf{x}) : \mathbf{x} \in \mathcal{B}_0 \rightarrow \mathbf{y} \in \mathcal{B}$, mapping points $\mathbf{x}$ in the reference configuration $\mathcal{B}_0$ to points $\mathbf{y}$ in the deformed configuration $\mathcal{B}$ and a concentration field, $c_H(\mathbf{x}) : \mathbf{x} \in \mathcal{B}_0 \rightarrow [0, 1]$, of the fraction of interstitial lattice positions occupied by hydrogen. The total free energy density of this system is composed of mechanical, chemical and gradient contributions:

$$f_{\text{total}} = f_{\text{mech}} + f_{\text{chem}} + f_{\text{grad}}. \tag{1}$$

The constitutive model for the mechanical free energy density is presented first. Here, the total deformation gradient, $\mathbf{F} = \operatorname{grad} \chi$, is multiplicatively decomposed into an elastic, chemical interstitial and plastic component as

$$\mathbf{F} = \mathbf{F}_e \mathbf{F}_i \mathbf{F}_p. \tag{2}$$

The elastic deformation gradient, $\mathbf{F}_e$, determines the stress at a material point, where an anisotropic elastic stiffness, $\mathbb{C}$, relates the elastic GREEN–LAGRANGE strain measure [25], $\mathbf{E}$, to the second PIOLA–KIRCHHOFF stress measure, $\mathbf{S}$ [25]:

$$\mathbf{E} = \frac{1}{2}(\mathbf{F}_e{}^{\mathsf{T}}\mathbf{F}_e - \mathbf{I}), \text{ and } \mathbf{S} = \mathbb{C}\,\mathbf{E}. \tag{3}$$

The chemical deformation gradient, $\mathbf{F}_i$, is determined from the hydrogen concentration, c_H,

$$\mathbf{F}_i = \epsilon_H c_H \mathbf{I} \tag{4}$$

and results from the interstitial volumetric change associated with solute hydrogen occupancy, ϵ_H. As we explain in more detail in Section 3.2, the concentration field c_H will be evaluated based on different equilibrium approximations. These distinguish whether no defects, dislocations or dislocations and grain boundaries are present in the system. The plastic deformation gradient evolves according to the flow rule

$$\dot{\mathbf{F}}_p = \mathbf{L}_p \mathbf{F}_p, \tag{5}$$

where the plastic velocity gradient, $\mathbf{L}_p$, is driven by the stress through the plasticity model. The crystal plasticity model used in the present study, is an adoption of the phenomenological description of Peirce et al. [26] for face-centered cubic crystals. The plastic velocity gradient $\mathbf{L}_p$ is composed of the slip rates $\dot{\gamma}^\alpha$ on each of the 12 BCC $\{110\}\langle 111\rangle$ slip systems, which are indexed by $\alpha = 1, \ldots, 12$.

$$\mathbf{L}_p = \sum_\alpha \dot{\gamma}^\alpha \, \mathbf{b}^\alpha \otimes \mathbf{n}^\alpha, \tag{6}$$

where $\mathbf{b}^\alpha$ and $\mathbf{n}^\alpha$ are unit vectors along the slip direction and slip plane normal, respectively. The slip rates are given by

$$\dot{\gamma}^\alpha = \dot{\gamma}_0 \left| \frac{\tau^\alpha}{g^\alpha} \right|^n \operatorname{sgn}(\tau^\alpha) \tag{7}$$

in terms of the resolved shear stress, $\tau^\alpha = \mathbf{S} \cdot (\mathbf{b}^\alpha \otimes \mathbf{n}^\alpha)$. The slip resistances on each slip system, g^α, evolve asymptotically towards g_∞ with shear γ^β ($\beta = 1, \ldots, 12$) according to the relationship

$$\dot{g}^\alpha = \dot{\gamma}^\beta h_0 \left| 1 - g^\beta/g_\infty \right|^a \operatorname{sgn}\left(1 - g^\beta/g_\infty \right) h_{\alpha\beta} \tag{8}$$

with parameters h_0 and a. The interaction between different slip systems is captured by the hardening matrix $h_{\alpha\beta}$.

The constitutive model represents an implicit system of equations to be solved for a consistent elastic and plastic deformation gradient (for details see [27]). The mechanical free energy density is then given by

$$f_{mech} = \frac{1}{2} \mathbf{S} \cdot \mathbf{E}$$ (9)

and its minimization results in mechanical equilibrium in terms of the first Piola–Kirchhoff stress measure, $\mathbf{P}$,

$$\frac{\delta f_{mech}}{\delta \chi} = \mathrm{Div}\, \mathbf{P} = \mathbf{0}.$$ (10)

The chemical free energy density for this system is based on the regular solution model

$$f_{chem} = \frac{1}{\Omega} \Big[E_H c_H + k_B T [c_H \ln c_H + (1 - c_H) \ln(1 - c_H)] \Big],$$ (11)

where Ω is the atomic volume, k_B is the BOLTZMANN constant, T is the temperature and E_H is a hydrogen enthalpy. Though E_H belongs to bulk Fe, the development of the hydrogen energy description given in the subsequent part of this publication will also include dislocations and grain boundaries.

Following [28], the free energy of the interface is given by

$$f_{grad} = \kappa |\mathrm{Grad}\, c_H|^2,$$ (12)

where κ is the surface energy parameter associated with the diffuse pore–matrix interface. The evolution of the conserved concentration field is then given by the modified Cahn–Hilliard equation

$$\dot{c}_H = \mathrm{Grad} \cdot M_H \, \mathrm{Grad}\, \mu_H,$$ (13)

where M_H is the mobility of the hydrogen solute. Their chemical potential μ_H is thermodynamically determined from the free energy density

$$\mu_H = \frac{\delta f_{total}}{\delta c_H} = \frac{\delta f_{mech}}{\delta c_H} + \frac{E_H}{\Omega} + \frac{k_B T}{\Omega} \ln \left(\frac{c_H}{1 - c_H} \right),$$ (14)

where the mechanical coupling with μ_H is obtained through Equation (4).

3.2. Numerical Implementation of the Continuum Model

For the numerical implementation of the chemoelastoplastic model, the Düsseldorf Advanced Material Simulation Kit (*DAMASK*) [22] is used, an open source crystal plasticity finite element library. For the sake of simplicity, we exemplarily introduce here representative volume elements (RVEs) used for simulations of fcc materials. The respective numerical parameters will be described correspondingly in the results part.

RVEs belong to the group of statistical approaches to achieve a representative homogenization of a material's microstructure for a macroscopic model. For plasticity models, the varying quantity typically is the dislocation density. In Figure 2, an example of the RVE for a dislocation density of $\rho = 1 \times 10^{15}$ m^{-2} is shown. The dislocation density ρ is given as $\rho = Nd/V$, where N is the number of dislocations, d the circumference of the dislocation line, modeled as ellipse, and V the volume of the simulation box. The semi major-axis is labelled a, the semi minor-axis b. The dimensions of the dislocation lines are $a = 15$ nm ± 3 nm and $b = 7$ nm ± 3 nm, where ± 3 nm here means that, for each individual dislocation line, a value from the interval $(-3 : 3)$ is randomly added to the corresponding basic value of a and b. The dislocations are equally distributed on all 12 slip systems, i.e., on all 12 slip planes, only one dislocation is initially set, and the two additional dislocations are randomly distributed to a slip system. These dislocations are randomly positioned on the corresponding slip systems. Such an RVE thus includes 25 different microstructural phases, i.e., 12 phases enclosed by the dislocations, the 12 dislocations and the matrix phase.

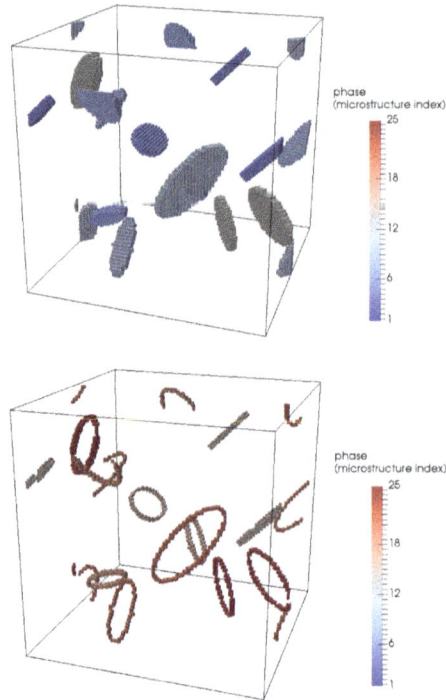

Figure 2. Here, the inner of a dislocation and the dislocation line are shown separately. top: the microstructure index 2–13 refers to the inner of dislocations, bottom: the microstructure index 14–25 refers to dislocation loops. Microstructure index 1 refers to the matrix and is not shown.

3.3. Atomistic Parameterisation

A key aim of the present work has been to determine the parameters of the continuum model from atomistic simulation. One important parameter is the effective binding energy of H in the vicinity of a dislocation. Evaluating such a binding energy is challenging for a number of reasons. Firstly, the long-range strain field around an ideal straight-line dislocation leads to the requirement of large simulation cells to avoid elastic artifacts. Moreover, care must be taken to avoid interactions between the interstitial H and its periodic images in the direction of the dislocation line. A third issue is that a large number of possible configurations must be considered in order to obtain reliable sampling of the potential energy surfce for hydrogen around such a dislocation.

In order to solve these issues, we utilise the environmental tight-binding (ETB) approach [29,30], which enables the rapid evaluation of energies and forces for systems of arbitrary chemistry, within a quantum-mechanical framework. The current approach requires the full eigenvalue spectrum of a one-electron Hamiltonian matrix, thus solving the one-electron Schrödinger equation within a local atomic-orbital basis set. Due to the cubic scaling of the eigenvalue problem, the present implementation is limited to systems of the order of 10^3 atoms. The reliability of the approach for the Fe-H system has been demonstrated by examining the H segregation behaviour at a selection of grain boundaries in α-Fe [30].

The chosen dislocation is the $a/2\langle 111\rangle$ screw dislocation in α-Fe, the behaviour of which is highly relevant for plastic deformation of iron at low temperatures. In order to circumvent the problem of long-range strain fields around a dislocation, and to avoid electronic surface effects arising from the application of elastic boundary conditions, the well-known *quadrupole* construction [31] is used, which allows for the study of dislocations within a periodic cell. In this approach, sketched schematically in Figure 3, a periodic arrangement of dislocations of opposite Burgers vectors is made, such that the long-range strain field is eliminated. One must take care that the distance between dislocation pairs is sufficiently large so that their mutual interaction (which is of course long-ranged) does not overshadow the effect under scrutiny. In these simulations, a rectangular arrangement of periodic dislocations is chosen, which allows for the computationally convenient use of an orthorhombic unit cell in the simulations; we found that a cell of 672 atoms is necessary to avoid significant augmentation of the core structures of the dislocations.

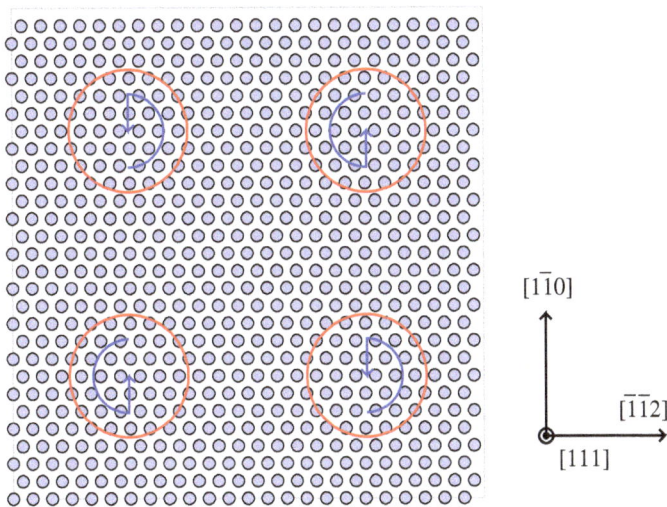

Figure 3. Schematic images of the quadrupole dislocation construction used in the simulation. The blue arrows indicate the helicity of the dislocations, with "up" and "down" arrows corresponding to screw dislocations of Burgers vectors $\mathbf{b} = a/2\langle 111\rangle$ and $-\mathbf{b}$, respectively. The red circles indicate the effective binding range.

Having obtained the relaxed dislocation configuration, the next stage is to assess the binding energy of hydrogen to this dislocation. In order to avoid interactions between H atoms and their periodic images, the simulation cell is extended by doubling the number of atomic planes along the direction of the Burgers vector, thus resulting in a H-free simulation cell of 1344 atoms. By numerical simulation, we find that the dislocation has an effective binding range of $\sim$5 Å from the centre of the core. Within this radius, 72 plausible interstitial sites for H are found (per Burgers vector), and the binding energy of each of these is evaluated (see Figure 4). The maximum value of the binding energy (with respect to a single H atom in a tetrahedral site in bulk α-Fe) is found to be 0.34 eV at the core of the dislocation. The median value of the binding energies is $\bar{E}_{\text{bind}} = 0.14$ eV, for a core radius $r_c = 5$ Å. To account for the entropic contribution of the dislocations, $\bar{E}_{\text{bind}}$ is shifted by $k_B T \ln(V_{dislocation}/V_{voxel})$ in the scale bridging simulations.

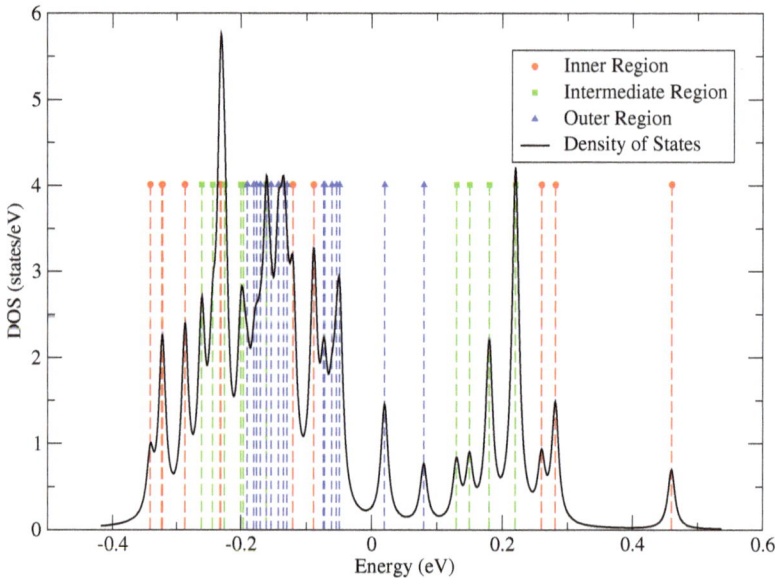

Figure 4. Each vertical line corresponds to one data point (due to symmetry, some points overlap). The colors correspond to the distances from the dislocation core, measured by the radius R. The red circles correspond to the 24 points closest to the dislocation core (R < 2.87 Å), the green squares are in the "intermediate range" (2.87 Å < R < 4.04 Å), with the blue points belong are the outermost 24 points. The density of states (DOS) curve is generated by broadening the delta functions by a Lorentzian function of width $k_B T/2$ (T = 300 K), with the integral under the DOS curve being normalised to 1.

While it would be possible to resolve the detailed picture presented in Figure 4 on a continuum level when we stick to small simulation volumina or low defect complexities, see the results reported [20,21], reaching process-relevant simulation boxes in the CPFEM approach requires further homogenziation. Therefore, to make a connection to the continuum picture, we choose the binding energies over a representative volume, such that the average H concentration within that volume corresponds to the expected density of occupied interstitial sites with respect to the bulk H chemical potential. The occupation of all interstitial sites within the dislocation line region will be one-half; this choice corresponds to selecting the *median* value of the binding energy distribution function. We note that, from the results obtained in terms of such comprisingly homogenised models, it is still possible to reconstruct the distinct site occupations, but this is restricted to fully relaxed states.

For the parametrisation of grain boundaries, we refer to the results published in [15] for α Fe. They show the dependence of the binding energy of hydrogen at an interstitial site on the spatial distance to the grain boundary. Precisely, we refer to the calculations for a Σ3 [1$\bar{1}$0] (112) grain boundary. The resulting binding energy shows a pronounced attraction with binding energies from 150 to 330 meV in a distance of about 2 Angstroms, while in distances from 2 to 7 Angstroms from the grain boundary plane, the binding energy ranges from 10 to 40 meV.

4. Results and Discussion

Here, the results of our investigations are presented and it is discussed how they can be interpreted in the context of grain boundary hydrogen embrittlement. On the phenomenological side, our main interest is in the competition between dislocation and grain boundaries in the aggregation of hydrogen. On the methodological side, the main goal is a quantitative scale transfer from the quantum mechanical description of hydrogen at dislocations and grain boundaries to an elastoplastic deformation model on the macroscale that reflects hydrogen aggregation. Apparently, one of the core challenges in such approaches is the reduction of complexity of the model while still catching the essential effects. We reduce the complexity step by step, which allows us to consider in each stage of simplification the loss of accuracy we inevitably tolerate in the description.

4.1. Modelling Hydrogen Aggregation Considering Dislocation Effects

We aim at the introduction of a composite model for the chemical potential and hydrogen concentration in the RVEs. This approach introduces an analytic approximation for the non-mechanical contributions to the hydrogen chemical potential, leading to a decrease of the computational expense of the simulations. Therefore, we begin with the comparison of the corresponding analytical approximations for the hydrogen distribution in volume elements with and without dislocations to results from analoguous simulations. These approximations are valid for a single volume element. The most basic approximation excludes also the influence of dislocations, leading to an average concentration profile

$$\langle c(\mu, E) \rangle = \frac{\exp\left(\frac{\mu - E}{k_B T}\right)}{1 + \exp\left(\frac{\mu - E}{k_B T}\right)} \tag{15}$$

with an average chemical potential $\mu = \langle \mu \rangle$ and formation energy $E = E^0$ for the bulk description. To include the dependence on the dislocation density in the model, the average concentration is split into bulk and dislocation contributions:

$$\langle c \rangle = c_{\text{bulk}} + v_{\text{dis}} \left(c_{\text{dis}} - c_{\text{bulk}} \right). \tag{16}$$

Here, v_{dis} is the volume fraction of the dislocation cores and v_{bulk} the volume fraction of the bulk, which are defined as $v_{\text{dis}} = \pi r_c^2 \rho$ and $v_{\text{bulk}} = 1 - v_{\text{dis}}$, whereby r_c describes the radius of the dislocation core and, therefore, πr_c^2 describes the cross-sectional area of the dislocation line. The system is described in a stationary state $\dot{c} = 0$, and mechanical contributions are neglected. Equality of the chemical potentials then provides expression for the concentration for the bulk and the dislocations as $c_{\text{bulk}} = \langle c(\langle \mu \rangle, E^0) \rangle$ and $c_{\text{dis}} = \langle c(\langle \mu \rangle, E_{dis}) \rangle$ where we can, based on the calculations in Section 3.3, set $E_{\text{dis}} = \bar{E}_{\text{bind}}$.

When comparing the predictions based on the composite analytical model, Equation (15), to the full field simulation results in Figure 5, we see that just the inclusion of the averaged dislocation densities is a strict requirement for the accuracy of the approximation. While the performance of the analytical approximation is convincing when dislocations are included, the situation becomes much more complex as soon as we introduce grain boundaries in the system to simulate polycrystalline samples.

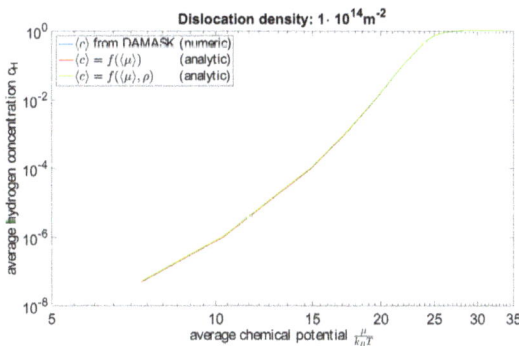

Figure 5. *Cont.*

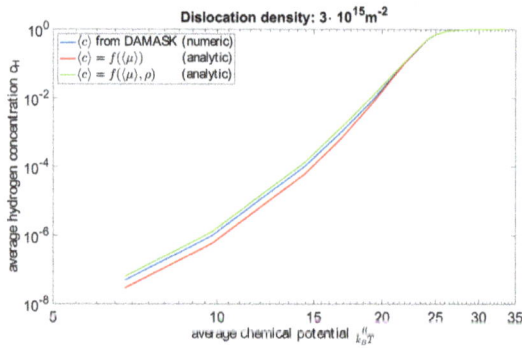

Figure 5. Comparison of the numerical solution (blue) with the old analytical solution (red), which includes no dislocation density and the new analytical solution (green) for different dislocation densities. Including the dislocation density substantially improves the accuracy of the analytically approximation.

4.2. Modelling Hydrogen Aggregation Considering Dislocation and Grain Boundary Effects

We distinguish two contributions to the hydrogen aggregation at grain boundaries here: on the one side, the stress concentration at the grain boundaries when the system is subjected to mechanical load and, on the other hand, the binding energies and binding length scales to the grain boundaries when the system is free from external stresses. For the latter, we introduce a voxel averaging scheme that allows us to include results from the ab initio calculations that are reported in [15] (see Figure 6).

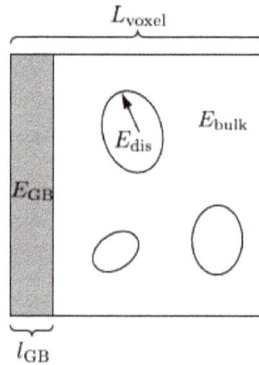

Figure 6. Illustration of grain boundary and dislocation volume voxel averaging.

Here, L_{voxel} is the length of the voxel and l_{GB} the range of attractive sites associated with the grain boundary. Then, the volume fraction of the grain boundary is defined as

$$v_{GB} = \frac{l_{GB}}{L_{voxel}}. \tag{17}$$

The average hydrogen concentration in this voxel can then be expressed as

$$\langle c \rangle = \langle c(\langle \mu \rangle, E^0)_{bulk} \rangle + v_{dis} \left(\langle c_{dis}(\langle \mu \rangle, E_{bind}) \rangle - (\langle c_{bulk}(\langle \mu \rangle, E^0) \rangle) \right) + v_{GB} \left(\langle c_{GB} \rangle - \langle c_{bulk}(\langle \mu \rangle, E^0) \rangle \right). \tag{18}$$

Via the ab initio results from [15], the concentration becomes dependent on the distance to the grain boundary as $E_{GB} = E_{GB}(r)$ when r is the distance from the grain boundary, thus

$$\langle c_{GB} \rangle = \frac{1}{l_{GB}} \int_0^{l_{GB}} \langle c_{GB}(r) \rangle \, dr. \tag{19}$$

However, the grain boundary fraction we introduce via this scheme is then fixed for a given voxel length, which leads to undesirable effects. When the voxel length is set to $L_{voxel} = 1$ μm, we obtain a hydrogen concentration profile as shown in Figure 7, corresponding to an estimated grain boundary fraction of ~10^{-3}. This scenario shows basically no remaining hydrogen segregation at the dislocations in the grains. To reduce the grain boundary fraction to values that show less dominant hydrogen aggregation at the grain boundaries, a voxel length of about 100 micrometres has to be chosen (see the simulation result obtained for that case in Figure 8). However, in that case, the numerical parametrisation implies a solution of the elastoplastic equations on an undesirably large spatial scale.

At this point, we have to recognise that defining a representative grain boundary binding energy, which can be used to obtain a composite model, i.e., analytical description, is a rather ineffective approach in comparison to the composite model for the dislocations. The spatial distribution of dislocations can be assumed to suffice the requirements to be fulfilled for a representative volume element definition, but the distribution of grain boundaries is subject to complex geometric constraints. Furthermore, the characteristic length scale of the grain boundary distribution is typically several orders of magnitude larger than the characteristic length scale of the dislocation distribution. This comparably small density requires larger representative volumes for the composite model, which would introduce further inaccuracies to the model.

To reflect the simulation results, we relate them to classical McLean [32] segregation profiles based on the same ab initio data sets. This includes estimates for the hydrogen segregation from dislocations to grain boundaries and the effect of hydrogen enrichment at the grain boundaries on segregation to further increased concentration levels.

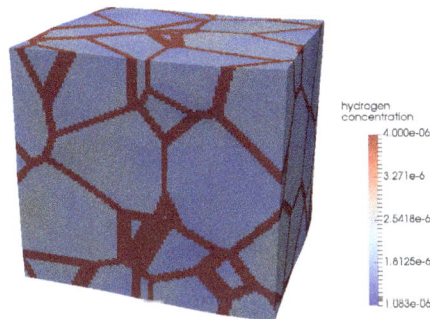

Figure 7. Simulation of a polycrystal containing dislocations with a binding energy to grain boundaries and a voxel size of $L_{voxel} = 1$ μm.

We estimate the influence of a locally hydrogen enriched grain boundary region on the segregation behaviour of additional hydrogen to those remaining attractive sites at the grain boundary. For this sake, we assume that the work of volume expansion due to the hydrogen is the dominant contribution to the solution energies. Furthermore, we assume that the change in the work of volume expansion under hydrogen solution with increasing hydrogen content is dominated by the change of the bulk modulus of the region at the grain boundary.

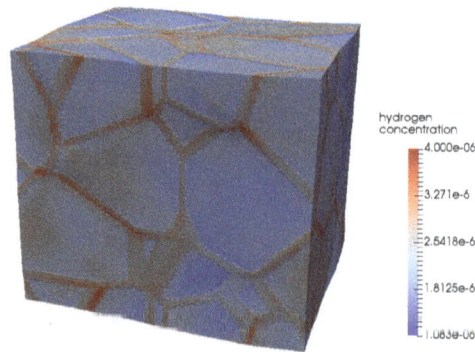

Figure 8. Simulation of a polycrystal containing dislocations with a binding energy to grain boundaries and a voxel size of $L_{voxel} = 100$ μm.

Therefore, we consider the following two limiting cases. First, when no other hydrogen atoms are present at the GB, the resulting segregation profile in a simple McLean picture is just based on the energies reported in [15]. In the second limiting case, we assume that half of all locally available, attractive sites, i.e., with a formation enthalpy ≤ 0.25 eV, are populated. For the $\Sigma_3[1\bar{1}0](112)$ bcc grain boundary, this corresponds to a minimum of three hydrogen atoms per volume $V \leq 10^{-28}$ m^3, which is in the order of 10% at hydrogen at the grain boundary. To estimate the difference in the segregation energy, assume $\Delta E_{el} = \nu \Delta B_0$ is assumed, with bulk modulus contrast ΔB_0 and partial molar volume of hydrogen ν. The exact elastic grain boundary data we need is not available, but the change of the bulk modulus due to the hydrogen aggregation at the grain boundary is estimated based on the bulk results reported in [33]. The resulting change ΔB_0 is approximated as 15 GPa. For the partial molar volume of hydrogen, we refer to the comprising studies summarized in [34], which suggest a constant value of 1.7×10^{-6} m^3/mol of atomic hydrogen (half of a H$_2$ molecule) over wide temperature and pressure ranges. The shift for the segregation energy then amounts to $\Delta E_{el} \approx 15 \times 1.7 \times 10^3/N_A$ eV/J, i.e., $\Delta E_{el} \approx 255$ meV.

This value certainly represents a grain boundary that is very densely populated by hydrogen. As the most attractive sites are restricted to a distance of 1–2 Angstroms away from the grain boundary, we assume that the elastic shift of the segregation energies only affects those sites that lie in this area. For the Σ_3 grain boundary investigated in [15], this corresponds to the four sites within approx. 2 Angstroms distance.

The resulting segregation profiles are shown in Figure 9, and the effect of the local elastic softening on the segregation profile is partially compensated when also dislocations as hydrogen traps are taken into account. While the detailed data presented in Figure 4 shows a maximal attraction of 0.34 eV in the area close to the dislocation core, we use here the averaged dislocation binding energy of 0.14 eV, corrected by an entropic contribution about $k_B T \ln (V_{dislocation}/V_{voxel})$. This is consistent as it reproduces an occupation of the dislocation-associated sites, which is reasonable for hydrogen enriched regions, and we make a similar assumption for the estimate of the occupation shift due to the hydrogen induced elastic softening. For an interpretation in the context of macroscopic metallurgical processes, a more detailed description of the ambient hydrogen chemical potential is required. Apart from the surface properties of the samples, especially surface roughness and surface porosity, the humidity of the atmosphere is also essential. When the samples are subjected to large thermal gradients due to heat treatment, the distribution of hydrogen at grain boundaries and dislocations close to the surface will change depending on the distance to the surface.

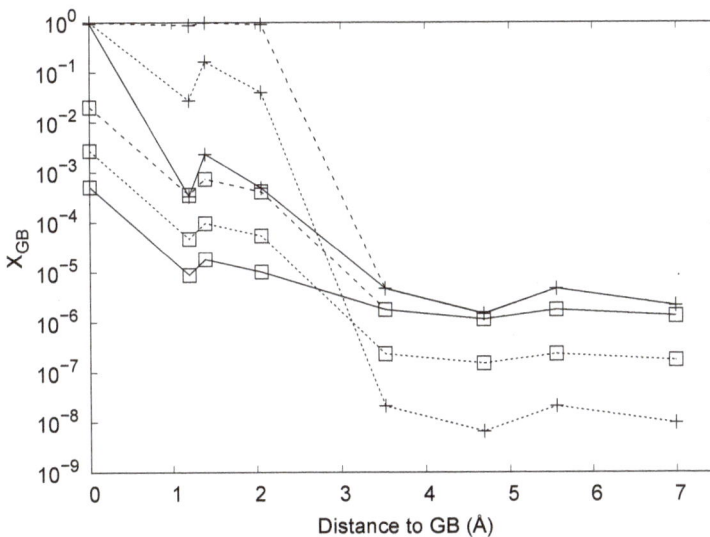

Figure 9. Segregation profiles for the sites in the vicinity of the grain boundary. The curves with cross symbols correspond to a temperature of 300 K, the curves with box symbols correspond to a temperature of 800 K. For both temperatures, the solid lines correspond to the segregation from a bcc bulk site with formation enthalpy $\Delta E = 0.25$ eV to the sites with energies and spatial distance to the grain boundary plane as reported in [15]. The dashed curves assume that the segregating hydrogen atom moves into a hydrogen enriched region which is limited to those sites within 2 Angstroms distance to the grain boundary plane. The hydrogen enriched zone has a reduced bulk modulus, and the resulting shift is about 255 meV. The dotted lines correspond to data that takes account that high local plasticity densities act as effective traps and that the hydrogen has to overcome the binding energy of about 140 meV. We note that the hydrogen is absorbed into the dislocations both for 300 K and 800 K from those sites that are more than 2 Angstroms away, expressed by positive segregation energies. This leads to a weak depopulation of this region at 800 K and a pronounced depopulation at 300 K.

This resulting effect of hydrogen binding to dislocations and hydrogen enrichment due to reduced mechanical resistance to hydrogen aggregation is a high local hydrogen density at the grain boundary. Though these results are based on estimates for the bulk modulus contrast and an effective dislocation binding energy, which result from a site occupation median, they exhibit the weakness of ferritic grain boundaries to hydrogen accumulation. For increasingly high levels of hydrogen aggregation to the grain boundary, as they are required for hydrogen embrittlement, a kinetic transport of hydrogen is required in addition to thermodynamically driven transport. As recently pointed out in [35], hydrogen shielded slip transfer to grain boundaries might offer not only a source for grain boundary stress concentration, but also this non-thermodynamical hydrogen transport process.

At this point, it is worth discussing the effect of the presence of hydrogen atoms along the grain boundaries on the plastic behavior on the polycrystalline metals. Recent simulations [35] show that hydrogen atoms have multiple effects of the dislocation–GB interactions. First of all, the segregated hydrogen atoms can develop stress field around the grain boundaries. These stress fields which stem from the misfit volumetric strain of the H atoms can attract/repel the dislocations. Thus, the average slip along the GBs can change [35]. This change promotes the accumulation of slip in local regions along the boundary that can lead to the formation of nano-cracks and voids.

Moreover, the presence of H atoms can not only significantly increase the critical shear stress needed for resolving the lattice dislocation in the grain boundaries, but it can also change the nature of the GB–dislocation interaction. The presence of H atoms can block the dissociation of the lattice dislocations into GB-dislocation. Thus, the slip either remains along the grain boundary or is transited to the adjacent grain at significantly higher stresses. This leads to the formation of more populated pile-ups and eventually leads to intergranular fracture of the grain boundary surfaces. The dislocations that are present in the pile ups can attract hydrogen atoms and deliver it to the grain boundary. As shown in previous studies in nickel, these extra hydrogen atoms can reduce the fracture energy significantly [36] and make brittle fracture favourable. This intergranular fracture cannot be achieved by considering only the equilibrium segregation hydrogen atoms along the boundary [37]. However, clarification of the kinetic aspects of this process needs further investigation.

5. Conclusions

Within our attempt to find efficient and predictive models for hydrogen segregation based on ab initio data, we present an analytic composite model for dislocation RVEs. Due to the comparably homogeneous distribution of dislocations, the length scales of defect separation and defect extent are unproblematic for the representative volume approach for dislocations and we can introduce an effective binding energy that leads to good agreement in the efficient composite model and full field simulations. When extended to also include grain boundaries in an RVE spirit, also based on ab initio determined binding energies, we recognise a conceptual difficulty of the analytical model. This challenge originates from the strong seperation of the length scales describing defect extent and defect distribution in the case of grain boundaries. A thorough approach to this problem would first demand a comprising analysis of the limitations of scale transfers between the description of grain boundaries on the ab initio scale and within the crystal plasticity picture, which models grain boundaries indirectly. Therefore, one of the key challenges is the vast combinatorical complexity of compositional and structural degrees of freedom in grain boundaries. On the one hand, the computational expense of state-of-the-art grain boundary calculations, which still exhibit substantial discrepancies to experimental measurements, is enormous. On the other hand, there are few approximative schemes available to interpolate or extrapolate data sets to differing grain boundaries, even if such data sets are available. Consequently, any comparability to experiments is limited to well defined model systems at the moment.

Apart from this methodological challenge, the simulations and theoretical considerations predict a strong hydrogen segregation in atomic distance from grain boundaries well in competition with dislocation based aggregation when local hydrogen occupation is elevated at the grain boundary. This competition between hydrogen segregation at dislocation and grain boundaries is also observed in the experimental measurements. In the low hydrogen concentration condition, the cleavage fracture indicates weak interaction between diffusive hydrogen and grain boundary, whereas the interaction becomes intensified with a higher amount of hydrogen. Therefore, there is still a challenge to quantitatively define the critical hydrogen contents leading to grain boundary decohesion for different material groups.

Author Contributions: All authors of the manuscript have contributed to the investigations that produced the presented results and the writing of the manuscript.

Acknowledgments: This work has been supported by the Collaborative Research Center 761 "Stahl ab initio" of the German Research Foundation. The authors gratefully acknowledge the computing time on the supercomputer JURECA at Forschungszentrum Jülich.

Conflicts of Interest: The authors declare no conflict of interest.

References

1. Petch, N.J. The lowering of fracture-stress due to surface adsorption. *Philos. Mag.* **1956**, *1*, 331–337. [CrossRef]
2. Birnbaum, H.K.; Sofronis, P. Hydrogen-enhanced localized plasticity—A mechanism for hydrogen-related fracture. *Mater. Sci. Eng. A* **1994**, *176*, 191–202. [CrossRef]
3. Kirchheim, R. Reducing grain boundary, dislocation line and vacancy formation energies by solute segregation. I. Theoretical background. *Acta Mater.* **2007**, *55*, 5129–5138. [CrossRef]

4. Kirchheim, R. Reducing grain boundary, dislocation line and vacancy formation energies by solute segregation: II. Experimental evidence and consequences. *Acta Mater.* **2007**, *55*, 5139–5148. [CrossRef]

5. Lynch, S. Environmentally assisted cracking: Overview of evidence for an adsorption-induced localised-slip process. *Acta Metall.* **1988**, *36*, 2639–2661. [CrossRef]

6. Nagumo, M. Hydrogen related failure of steels—A new aspect. *Mater. Sci. Technol.* **2004**, *20*, 940–950. [CrossRef]

7. Takai, K.; Shoda, H.; Suzuki, H.; Nagumo, M. Lattice defects dominating hydrogen-related failure of metals. *Acta Mater.* **2008**, *56*, 5158–5167. [CrossRef]

8. Crabtree, G.W.; Dresselhaus, M.S.; Buchanan, M.V. The hydrogen economy. *Phys. Today* **2004**, *57*, 39–44. [CrossRef]

9. Turner, J.A. Sustainable hydrogen production. *Science* **2004**, *305*, 972–974. [CrossRef] [PubMed]

10. Figueroa, D.; Robinson, M. The effects of sacrificial coatings on hydrogen embrittlement and re-embrittlement of ultra high strength steels. *Corros. Sci.* **2008**, *50*, 1066–1079. [CrossRef]

11. Rhodes, P. Environment-assisted cracking of corrosion-resistant alloys in oil and gas production environments: A review. *Corrosion* **2001**, *57*, 923–966. [CrossRef]

12. Barrera, O.; Bombac, D.; Chen, Y.; Daff, T.; Galindo-Nava, E.; Gong, P.; Haley, D.; Horton, R.; Katzarov, I.; Kermode, J.; et al. Understanding and mitigating hydrogen embrittlement of steels: A review of experimental, modelling and design progress from atomistic to continuum. *J. Mater. Sci.* **2018**, *53*, 6251–6290. [CrossRef]

13. Jemblie, L.; Olden, V.; Akselsen, O.M. A review of cohesive zone modelling as an approach for numerically assessing hydrogen embrittlement of steel structures. *Philos. Trans. R. Soc. A* **2017**, *375*, 20160411. [CrossRef] [PubMed]

14. Koyama, M.; Akiyama, E.; Lee, Y.K.; Raabe, D.; Tsuzaki, K. Overview of hydrogen embrittlement in high-Mn steels. *Int. J. Hydrogen Energy* **2017**, *42*, 12706–12723. [CrossRef]

15. Du, Y.A.; Ismer, L.; Rogal, J.; Hickel, T.; Neugebauer, J.; Drautz, R. First-principles study on the interaction of H interstitials with grain boundaries in α-and γ-Fe. *Phys. Rev. B* **2011**, *84*, 144121. [CrossRef]

16. Troiano, A.R. The role of hydrogen and other interstitials in the mechanical behavior of metals. *Trans. ASM* **1960**, *52*, 54–80. [CrossRef]

17. Oriani, R. A mechanistic theory of hydrogen embrittlement of steels. *Ber. Bunsengesellsch. Phys. Chem.* **1972**, *76*, 848–857.

18. Oriani, R.; Josephic, P. Equilibrium aspects of hydrogen-induced cracking of steels. *Acta Metall.* **1974**, *22*, 1065–1074. [CrossRef]

19. Baechem, C. A New Model for Hydrogen-Assisted Cracking. *Met. Trans.* **1972**, *3*, 437–451.

20. Von Pezold, J.; Lymperakis, L.; Neugebeauer, J. Hydrogen-enhanced local plasticity at dilute bulk H concentrations: The role of H–H interactions and the formation of local hydrides. *Acta Mater.* **2011**, *59*, 2969–2980. [CrossRef]

21. Leyson, G.; Grabowski, B.; Neugebauer, J. Multiscale description of dislocation induced nano-hydrides. *Acta Mater.* **2015**, *89*, 50–59. [CrossRef]

22. Roters, F.; Eisenlohr, P.; Kords, C.; Tjahjanto, D.; Diehl, M.; Raabe, D. DAMASK: The Duesseldorf Advanced Material Simulation Kit for studying crystal plasticity using an FE based or a spectral numerical solver. *Procedia IUTAM* **2012**, *3*, 3–10. [CrossRef]

23. Hirth, J. Effects of hydrogen on the properties of iron and steel. *Metall. Trans. A* **1980**, *11*, 861–890. [CrossRef]

24. Shanthraj, P.; Sharma, L.; Svendsen, B.; Roters, F.; Raabe, D. A phase field model for damage in elasto-viscoplastic materials. *Comput. Methods Appl. Mech. Eng.* **2016**, *312*, 167–185. [CrossRef]

25. Nemat-Nasser, S.; Hori, M. *Micromechanics: Overall Properties of Heterogenous Materials*; Elsevier North-Holland: Amsterdam, The Netherlands, 1999.

26. Peirce, D.; Asaro, R.; Needleman, A. Material rate dependence and localized deformation in crystalline solids. *Acta Metall.* **1983**, *31*, 1951–1976. [CrossRef]

27. Roters, F.; Eisenlohr, P.; Hantcherli, L.; Tjahjanto, D.; Bieler, T.; Raabe, D. Overview of constitutive laws, kinematics, homogenization and multiscale methods in crystal plasticity finite-element modeling: Theory, experiments, applications. *Acta Mater.* **2010**, *58*, 1152–1211. [CrossRef]

28. Cahn, J.W.; Hilliard, J.E. Free energy of a nonuniform system. I. Interfacial free energy. *J. Chem. Phys.* **1958**, *28*, 258–267. [CrossRef]

29. McEniry, E.J.; Drautz, R.; Madsen, G. Environmental tight-binding modeling of nickel and cobalt clusters. *J. Phys. Condens. Matter* **2013**, *25*, 115502. [CrossRef] [PubMed]

30. McEniry, E.J.; Hickel, T.; Neugebauer, J. Hydrogen behaviour at twist {110} grain boundaries in α-Fe. *Philos. Trans. A* **2017**, *375*, 20160402. [CrossRef] [PubMed]

31. Clouet, E.; Ventelon, L.; Willaime, F. Dislocation core field. II. Screw dislocation in iron. *Phys. Rev. B* **2011**, *84*, 224107. [CrossRef]

32. Lejcek, P. *Grain Boundary Segregation in Metals*; Springer: Berlin, Germany, 2010.

33. Psiachos, D.; Hammerschmidt, T.; Drautz, R. Ab initio study of the modification of elastic properties of alpha-iron by hydrostatic strain and by hydrogen interstitials. *Acta Mater.* **2011**, *59*, 4255–4263. [CrossRef]

34. Peisl, H. *Topics in Applied Physics: Hydrogen in Metals I*; Springer: Berlin, Germany, 1978.

35. Tehranchi, A. *Atomistic Mechanisms of Hydrogen Embrittlement*; EPFL: Lausanne, Switzerland, 2017; p. 165.

36. Tehranchi, A.; Curtin, W. Atomistic study of hydrogen embrittlement of grain boundaries in nickel: II. Decohesion. *Model. Simul. Mater. Sci. Eng.* **2017**, *25*, 075013. [CrossRef]

37. Tehranchi, A.; Curtin, W. Atomistic study of hydrogen embrittlement of grain boundaries in nickel: I. Fracture. *J. Mech. Phys. Solids* **2017**, *101*, 150–165. [CrossRef]

metals

MDPI

Article

Stability, Electronic Structure, and Dehydrogenation Properties of Pristine and Doped 2D MgH$_2$ by the First Principles Study

Xu Gong and Xiaohong Shao *

College of Science, Beijing University of Chemical Technology, Beijing 100029, China;
2016200908@mail.buct.edu.cn
* Correspondence: shaoxh@mail.buct.edu.cn; Tel.: +86-10-64433867

Received: 30 May 2018; Accepted: 18 June 2018; Published: 25 June 2018

Abstract: Based on first principles calculations, we theoretically predict the new two-dimensional (2D) MgH$_2$. The thermodynamic stability, partial density of states, electron localization function, and Bader charge of pure and the transition metal (Ti, V, and Mn) doped 2D MgH$_2$ are investigated. The results show that all the systems are dynamically stable, and the dehydrogenation properties indicate that the decomposition temperature can be reduced by introducing the transition metal, and the Mn doped system exhibits good performance for better hydrogen storage and dehydrogenation kinetics.

Keywords: 2D MgH$_2$; hydrogen storage; first principles; dehydrogenation kinetics

---

1. Introduction

Hydrogen energy is considered to be the most promising alternative because it is lightweight, environmentally friendly, highly efficient, renewable, and abundant on earth. However, the storage limits the application of hydrogen. Metal hydrides are considered as the most promising materials for hydrogen storage and have been widely investigated in the past decades [1]. Among them, magnesium-based alloys and magnesium hydrides can achieve the hydrogen storage capacity of 7.6 wt % [2–8]. However, the high thermodynamic stability (the heat of formation is around −75.99 kJ/mol·H$_2$), high desorption temperatures (above 573 K), and slow dehydrogenation kinetics seriously limit the practical applications [3,9,10]. Therefore, it is always a central task to design new materials or adopt efficient strategies for achieving lower desorption temperatures and good dehydrogenation performances.

Previous studies show that the bonding nature of MgH$_2$ is a mixture of strong ionic and weak covalent bonding [11], and weakening the interactions may be an effective strategy to improve dehydrogenation performance. It has been reported that doping with transition metal elements or their oxides mixtures with MgH$_2$ can effectively reduce its stability and improve the hydrogen desorption thermodynamics [3,12–17]. Oelerich [15] et al. have reported that MgH$_2$ milled with Fe$_3$O$_4$, V$_2$O$_5$, Mn$_2$O$_3$, or Cr$_2$O$_3$, etc. can accelerate the hydrogen desorption kinetics. Shang [3] et al. have studied the hydrogen storage performance of (MgH$_2$ + M) systems (M = Al, Ti, Fe, Ni, Cu, and Nb) experimentally and theoretically, and they found that MgH$_2$ mixed with those metals can reduce the stability and improve the hydrogen desorption kinetics. Nonetheless, the MgH$_2$ systems still have a high desorption temperature around 500 K. It is noted that the bulk MgH$_2$ has been extensively investigated, however, the single-layer magnesium hydrides have been largely ignored. Motivated by the above mentioned details, we focus on exploring new structures with good dehydrogenation performance in this work.

In this paper, the new two-dimensional (2D) MgH$_2$ structure is theoretically predicted and studied by first principles calculations. The stabilities of pure and Ti/V/Mn doped MgH$_2$ are discussed by the phonon spectra and heat of formation. The calculated heat of formation for pure and Ti/V/Mn doped

2D MgH_2 are -37.57, -25.67, -18.14, and -23.90 kJ/mol·H_2, respectively, which are significantly lower than that of -75.99 kJ/mol·H_2 of bulk MgH_2. The electronic structure and hydrogen desorption kinetics results show that the predicted two-dimensional magnesium hydride are promising candidates for hydrogen storage.

2. Computational Details

The structural optimization and electronic property calculations were performed using the projector augmented plane-wave method (PAW) based on the density functional theory (DFT) in the Vienna ab initio simulation package (VASP) [18,19]. The exchange-correlation potential was approximated by generalized gradient approximation (GGA) in the Perdew-Burke-Ernzerhof (PBE) form [20,21]. To avoid the interlayer effects of the c-axis, the vacuum region around 15 Å was set in all the systems. The energy cutoff of 600 eV and the $9 \times 9 \times 1$ Γ-centered Monkhorst-Pack k-points [22] were employed for all calculations. The atomic positions were fully relaxed and the force tolerance between each atom was less than 0.01 eV/Å for the structural optimization. The convergence criteria of 10^{-6} eV per atom was applied to be self-consistent. Meanwhile, for calculation of electronic structures, we also applied the local density approximation (LDA) [23] and HSE06 [24] was functional. The kinetic stability was discussed using the phonon spectra calculations in PHONOPY code coupled with VASP using the density functional perturbation theory (DFPT) method [25–27].

3. Results and Discussion

Figure 1 shows the fully relaxed structure of the top and side view of pure 2D MgH_2 of the hexagonal structure with space group *P-3m1* (D_{3d}^3). The primitive cell has the lattice constant of $a = b = 3.01$ Å, the Mg-H bond length of $l = 1.97$ Å, and the buckled height of $d = 1.86$ Å. The next calculations were performed for the $3 \times 3 \times 1$ supercell of 2D MgH_2, named Mg_9H_{18}. The corresponding lattice parameters, Wyckoff [28] and atomic positions, are shown in Table 1. As is seen, there are nine Mg atoms located at 1b (Mg1), 6h (Mg2), and 2d (Mg3) sites, while the eighteen H atoms are located in three identical Wyckoff positions, i.e., 6i, as shown in Figure 1.

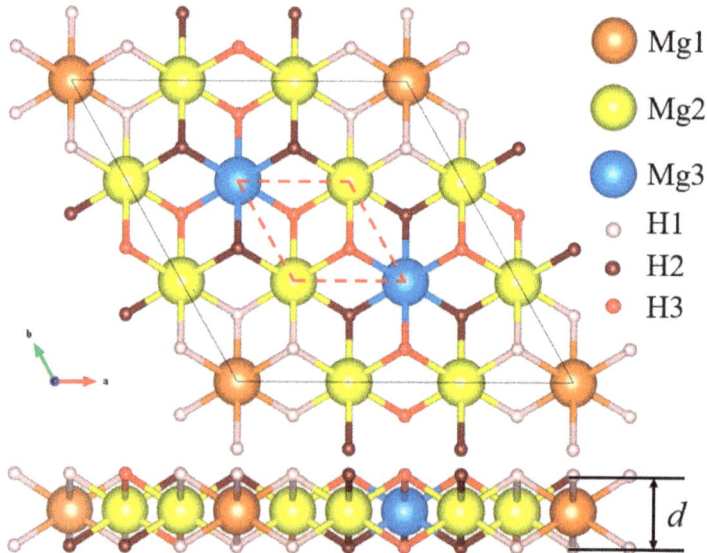

Figure 1. The relaxed unit cell of Mg_9H_{18}. The primate cell is marked with a red dashed box.

Table 1. The relaxed structural parameters and atomic positions of Mg_9H_{18}.

Lattice Parameters	Atom	Wyckoff Positions	Atomic Positions (Fractional)		
			x	y	z
164(*P-3m1*)	Mg1	1b	0	0	0.5
a = *b* = 9.033 Å	Mg2	6h	0	0.33333	0.5
c = 15 Å	Mg3	2d	0.33333	0.66667	0.5
d = 1.86 Å	H1	6i	0.11111	0.22222	0.43783
α = *β* = 90°	H2	6i	0.22222	0.44444	0.56217
γ = 120°	H3	6i	0.11111	0.55556	0.43783

In this work, three different Mg sites are considered as possible positions for substitution doping. Meanwhile, defects are inevitable in synthesis or processing and can usually affect their properties [29–33]. The most common types of defect are vacancy defects, so we also considered the vacancies of Mg (Mg_8H_{18}) for comparison to the doped systems. The formation energies were calculated to determine the favorable positions of doping elements of Ti/V/Mn, which is defined as $\Delta E = E_{tot}(Mg_8H_{18}X_n) - E_{tot}(Mg_9H_{18}) - n\,E_{tot}(X) + E_{tot}(Mg)$, where E_{tot} is the total energy of the system, the parameter $n = 0/1$ represents Mg vacancy, and X (X = Ti, V, and Mn) doped. The energies are listed in Table 2. It is noticed that the Mg_8H_{18} and Ti/V/Mn doped systems have positive energy, indicating that the stability of all the systems are lower than that of pure Mg_9H_{18}. In addition, for the three high symmetry sites of Mg1 (1b), Mg2 (6h), and Mg3 (2d), the ΔE are nearly identical, therefore, we assume that all the doped-sites are located at the Mg1 site in the following work. The relaxed parameters and bond lengths of Mg_9H_{18} and $Mg_8H_{18}X$ (X = Ti, V, and Mn) are listed in Table 2, and for the detailed lattice parameters, see Table A1 (Appendix A). As is seen, the bond length of Mg-H is changed, which indicates that the doped X atoms break the symmetry of the 2D MgH_2 structure.

Table 2. The energy (ΔE), the lattice parameter (*a*), and bond length of Mg_9H_{18}, Mg_8H_{18} and $Mg_8H_{18}X$ (X = Ti, V, and Mn).

Hydride	ΔE (eV)			Parameter	Bond Length (Å)					
	Mg1	Mg2	Mg3	a (Å)	Sub-H1	Mg2-H1	Mg2-H2	Mg2-H3	Mg3-H2	Mg3-H3
Mg_9H_{18}	0	0	0	9.033	1.972	1.972	1.972	1.972	1.972	1.972
Mg_8H_{18}	2.968	2.968	2.968	9.062	-	1.894	2.043	1.976	1.947	1.992
$Mg_8H_{18}Ti$	1.113	1.114	1.114	9.027	1.915	1.997	1.964	1.982	1.946	1.990
$Mg_8H_{18}V$	1.818	1.818	1.819	8.951	1.822	1.999	1.945	1.992	1.939	1.983
$Mg_8H_{18}Mn$	1.279	1.279	1.279	8.815	1.691	2.028	1.911	2.008	1.937	1.965

Structural stability is discussed by the phonon spectra calculations using the DFPT method, as is shown in Figure 2. Clearly, there are no imaginary frequencies in the whole Brillouin zone, indicating that all the systems are dynamically stable. Meanwhile, the heat of formation (ΔH) [7,34–36] is one of the most fundamentally thermodynamic properties. The heat of formation can be obtained directly from the equation $\Delta H = [E_{tot}(Mg_{9-n}H_{18}X_{n+m}) - (n + m)\,E_{tot}(X) - (9-n)\,E_{tot}(Mg) - 9\,E_{tot}(H_2)]/9$, where the parameters ($n = 0$, $m = 0$), ($n = 1$, $m = -1$), and ($n = 1$, $m = 0$), represent pure, Mg vacancy, and X (X = Ti, V and Mn) doped Mg_9H_{18}, respectively. The value of $E_{tot}(H_2)$ of -6.762 eV in a $10 \times 10 \times 10$ Å^3 cubic cell is very close to -6.773 eV reported in Ref. [37].

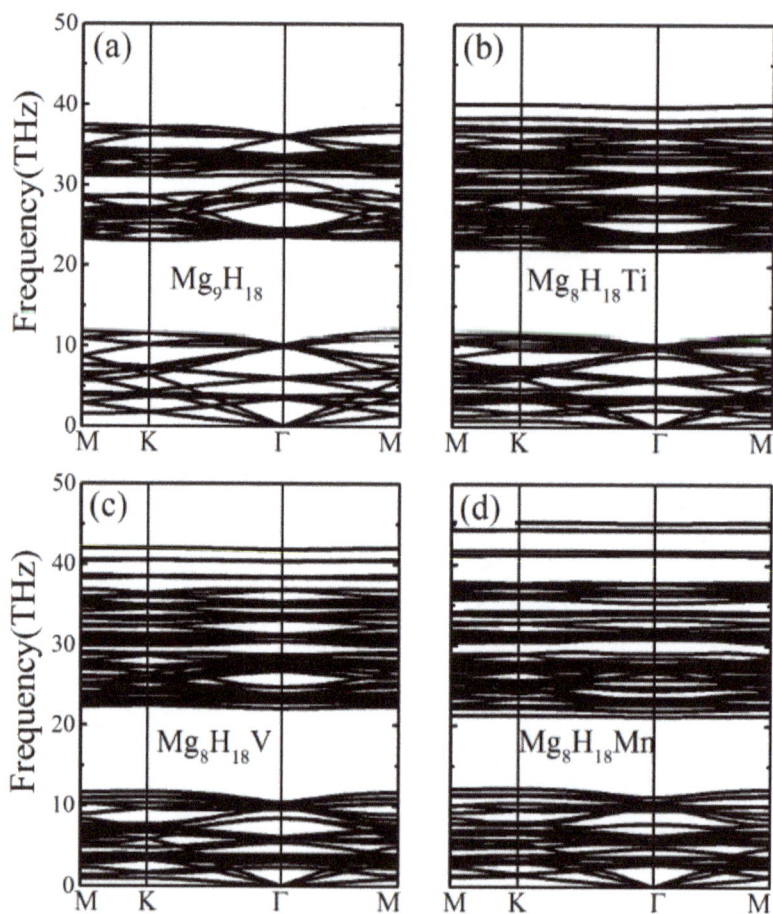

Figure 2. The phonon spectra of Mg_9H_{18} (**a**); $Mg_8H_{18}Ti$ (**b**); $Mg_8H_{18}V$ (**c**); and $Mg_8H_{18}Mn$ (**d**).

The estimated heats of formation are listed in Table 3. As is seen, the heat of formation of Mg_9H_{18}, Mg_8H_{18}, $Mg_8H_{18}Ti$, $Mg_8H_{18}V$, and $Mg_8H_{18}Mn$ are -37.57, 31.71, -25.67, -18.14, and -23.90 kJ/mol·H$_2$, respectively. The results show that the stability decreased for the doped 2D MgH_2, followed by $Mg_8H_{18}Ti$, $Mg_8H_{18}Mn$, and $Mg_8H_{18}V$, and Mg_8H_{18} is the most unstable. In comparison, we also obtained the heat of formation of the bulk MgH_2 of $\Delta H = -54.56$ kJ/mol·H$_2$, which is close to the theoretical values -54.4 in Ref. [36] and -53.85 kJ/mol·H$_2$ in Ref. [38]. At the same time, we estimated the decomposition temperature according to the following relationship: $\ln \frac{P}{P_0} = \frac{\Delta H}{RT} - \frac{\Delta S}{R}$, where P, P_0, R, T, and ΔS represent the pressure, the standard pressure, the gas constant, the decomposition temperature, and the entropy change, respectively. At the standard pressure, the ΔH is defined as $\Delta H = T\Delta S$ [39,40]. For most of the dehydrogenation reactions of simple metal hydrides, the ΔS is in the range of 95 J/mol·K $< \Delta S(H_2) < 140$ J/mol·K [41]. Consequently, the decomposition temperatures are 268 K $< T(Mg_9H_{18}) < 396$ K, 183 K $< T(Mg_8H_{18}Ti) < 270$ K, 130 K $< T(Mg_8H_{18}V) < 191$ K, 171 K $< T(Mg_8H_{18}Mn) < 252$ K, which are significantly lower than that of 573~673 K of bulk MgH_2. The discussions mentioned above show that 2D MgH_2 has better dehydrogenation thermodynamic properties than that of bulk MgH_2, and doping with Ti, V, and Mn elements can reduce the stability and improve the dehydrogenation thermodynamics properties of 2D MgH_2.

Table 3. The heat of formation (ΔH), the decomposition temperature (T), Bader charge of Mg and H atoms, and the dehydrogenation energies (E_d) of Mg_9H_{18}, Mg_8H_{18}, and $Mg_8H_{18}X$ (X = Ti, V, and Mn).

Hydride	ΔH	T	Bader Charge (e)			E_d
	(kJ/mol·H$_2$)	(K)	Mg	X	H	(eV)
Mg_9H_{18}	−37.57	268~396	+2.000	-	−0.997	1.589
Mg_8H_{18}	31.71	-	+2.000	-	−0.886	−1.931
$Mg_8H_{18}Ti$	−25.67	183~270	+2.000	+1.825	−0.988	1.305
$Mg_8H_{18}V$	−18.14	130~191	+2.000	+1.523	−0.971	1.044
$Mg_8H_{18}Mn$	−23.90	171~252	+2.000	+0.975	−0.940	0.853

To understand the effect of Ti/V/Mn-doped 2D MgH$_2$ well, we analyzed the electronic structures. The band structures were obtained using PBE, LDA, and HSE06 functionals and are shown in Figure A1. It can be seen that the pure Mg_9H_{18} with the energy gap is 4.87 eV, which is smaller than the experimental values 5.16 eV [42] or 5.6 eV [43] of bulk MgH$_2$. For comparison, we found that the bandgaps using HSE06 functional are larger than those using PBE and LDA functionals, and the bandgaps using LDA functional are close to those of the PBE functional for pure and vacancies of Mg_9H_{18}. For the doped systems, there are few energy bands across the Fermi level due to the d orbitals of the dopants. Figure 3 shows the total and partial density of states (PDOS) of Mg_9H_{18} and Mg_8H_{18}, which are calculated using the PBE functional. We can see the stronger hybridization between H and Mg atoms near the Fermi level, which indicates the strong interaction between H and Mg atoms. For the $Mg_8H_{18}X$, the electronic structure is different from that of the pure Mg_9H_{18}, as is shown in Figure 4. We can see that the d orbitals of Ti/V/Mn are mainly located near the Fermi level in doped-2D MgH$_2$, and there are few H-s orbitals and states of Mg atoms at the Fermi level, which indicates that the interactions between Ti/V/Mn and H/Mg atoms are relatively weaker. Meanwhile, since the H-s orbitals are reduced at the Fermi level, the hybridization between H and Mg atoms is weaker compared to pure Mg_9H_{18}.

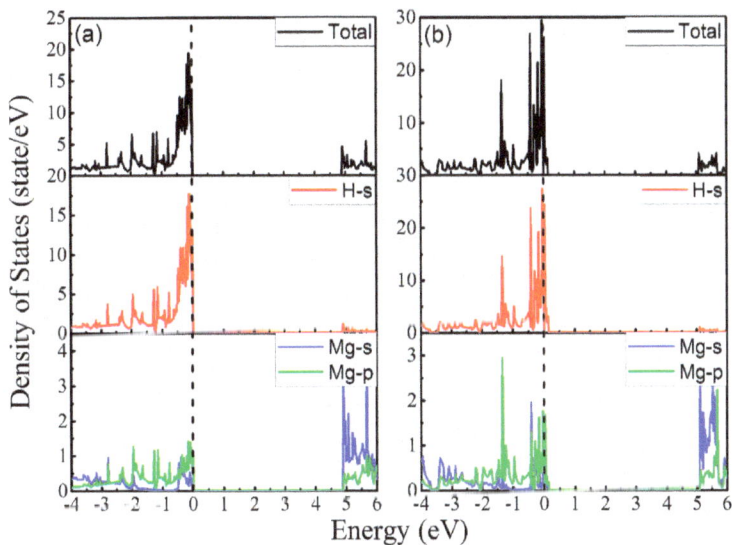

Figure 3. The total and partial densities of states of pure (**a**) and defective (**b**) Mg_9H_{18}.

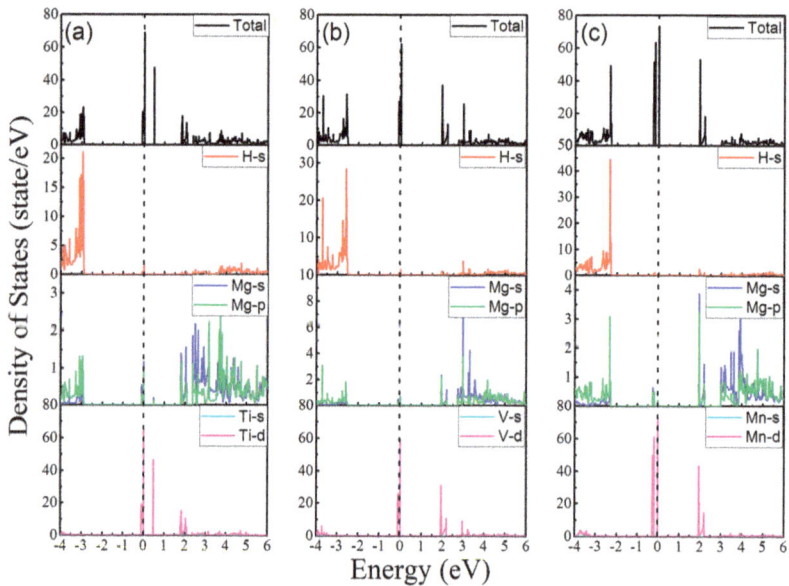

Figure 4. The total and partial densities of states of $Mg_8H_{18}Ti$ (**a**); $Mg_8H_{18}V$ (**b**); and $Mg_8H_{18}Mn$ (**c**).

In order to analyze the chemical bond of all the systems, the electron localization function (ELF) was calculated and is shown in Figure 5 with the isosurfaces of 0.6 e/$Å^3$. As shown in Figure 5b–f, all the systems have similar features in that the ELF values are lower between Mg and H atoms. These suggest that the ionic bonds exist between Mg and H atoms, and the Mg atoms act as the charge donor, according well with the discussions of PDOS and ELF of bulk MgH_2. The Bader charges were also calculated (see Table 3), and it can be seen that Mg atoms contributed two electrons, and H atoms acquired electrons to form anions. Meanwhile, the H atoms obtained fewer electrons due to the Mg vacancy and doping with Ti/V/Mn elements, thereby weakening the interactions between H and the metal atoms.

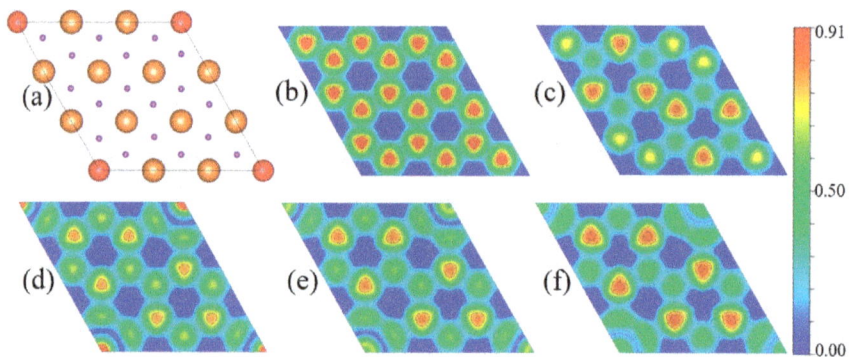

Figure 5. (**a**) Structural representation of considered systems. The big (small) ball represents Mg (H) and the red ball site is the doped site; (**b–f**) represent the electron localization function (ELF) of Mg_9H_{18}, Mg_8H_{18}, $Mg_8H_{18}Ti$, $Mg_8H_{18}V$, and $Mg_8H_{18}Mn$, respectively. The color bar represents the values of ELF.

As mentioned above, doping with Ti/V/Mn elements reduces the stability of 2D MgH_2 and weakens the interactions between H and metal atoms, which facilitates the release of hydrogen. To further understand the dehydrogenation behavior, the dehydrogenation energy was estimated by the formula: $E_d = E_{tot}(Mg_{9-n}H_{17}X_{n+m}) - E_{tot}(Mg_{9-n}H_{18}X_{n+m}) + 1/2\ E_{tot}(H_2)$, where ($n = 0$, $m = 0$), ($n = 1$, $m = -1$), and ($n = 1$, $m = 0$) represent the pure, Mg vacancy, and X (X = Ti, V, and Mn) doped Mg_9H_{18}, respectively. The dehydrogenation energies are listed in Table 3. The results show that the dehydrogenation energy of the Mg_8H_{18} was significantly reduced compared to the pure and doped Mg_9H_{18}, while there are high ΔE of 2.968 eV and positive ΔH of 31.71 kJ/mol·H_2, indicating that it is almost impossible to steadily occur. For doped systems, their dehydrogenation energies are significantly smaller than 1.589 eV of pure Mg_9H_{18}, especially $Mg_8H_{18}Mn$ with the dehydrogenation energy of 0.853 eV. Therefore, doping with Ti/V/Mn elements can improve the dehydrogenation thermodynamic properties of 2D MgH_2.

4. Conclusions

In summary, we theoretically predicted two-dimensional MgH_2 and studied the electronic and dehydrogenation properties of pure and Ti/VMn doped 2D MgH_2. The phonon spectra calculations indicate that all the systems are dynamically stable. The results of heat of formation suggests that Ti/V/Mn doping can reduce the thermodynamic stability, followed by $Mg_8H_{18}Ti$, $Mg_8H_{18}Mn$, and $Mg_8H_{18}V$, and Mg_8H_{18} is the most unstable. Importantly, the dehydrogenation temperatures for all the systems are significantly lower than that of bulk MgH_2 at 573~673 K. Especially, $Mg_8H_{18}Ti$ (183~270 K), $Mg_8H_{18}V$ (130~191 K), and $Mg_8H_{18}Mn$ (171~252 K) have much lower decomposition temperature than that of pure 2D MgH_2 (268~396 K), which is important for practical applications. The partial densities of states, electron localization function, and Bader charge calculation results show that Ti, V, and Mn elements can weaken the interaction between H and the metal atoms, which is favorable to dehydrogenation and better than that of the bulk MgH_2.

Author Contributions: All authors contributed equally to this work.

Acknowledgments: This work was supported by the National Natural Science Foundation of China (Grant No. 11604008) and by BUCT Fund for Disciplines Construction (Project No. XK1702).

Conflicts of Interest: The authors declare no conflicts of interest.

Appendix A

Table A1. The relaxed structure parameters of pure, vacancies, and doped Mg_9H_{18} using Perdew-Burke-Ernzerhof (PBE) and local density approximation (LDA) functionals.

Hydride	PBE				LDA			
	a (Å)	α (°)	β (°)	γ (°)	a (Å)	α (°)	β (°)	γ (°)
Mg_9H_{18}	9.033	90.0	90.0	120.0	8.894	90.0	90.0	120.0
Mg_8H_{18}	9.062	90.0	90.0	120.0	8.888	90.0	90.0	120.0
$Mg_8H_{18}Ti$	9.027	90.0	90.0	120.0	8.883	90.0	90.0	120.0
$Mg_8H_{18}V$	8.951	90.0	90.0	120.0	8.803	90.0	90.0	120.0
$Mg_8H_{18}Mn$	8.815	90.0	90.0	120.0	8.662	90.0	90.0	120.0

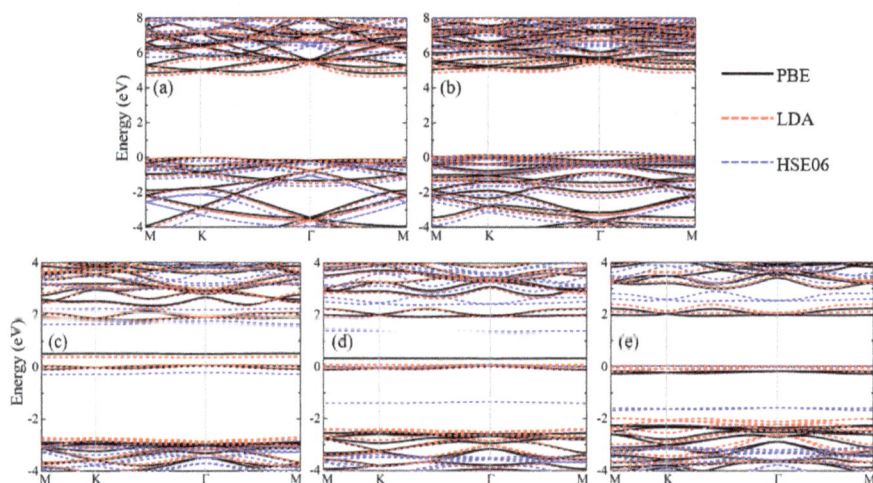

Figure A1. Band structures of Mg_9H_{18} (**a**); Mg_8H_{18} (**b**); $Mg_8H_{18}Ti$ (**c**); $Mg_8H_{18}V$ (**d**); and $Mg_8H_{18}Mn$ (**e**) calculated using PBE (black line), LDA (red dot line), and HSE06 (blue dot line) functionals, respectively.

References

1. Sakintuna, B.; Lamari-Darkrim, F.; Hirscher, M. Metal hydride materials for solid hydrogen storage: A review. *Int. J. Hydrogen Energy* **2007**, *32*, 1121–1140. [CrossRef]
2. Mohammed, Z.; Ahmed, R.; Mohammed Benali, K.; Bakhtiar ul, H.; Ahmad Radzi Mat, I.; Souraya, G.-S. First principle investigations of the physical properties of hydrogen-rich MgH_2. *Phys. Scr.* **2013**, *88*, 065704. [CrossRef]
3. Shang, C.X.; Bououdina, M.; Song, Y.; Guo, Z.X. Mechanical alloying and electronic simulations of $(MgH_2 + M)$ systems (M = Al, Ti, Fe, Ni, Cu and Nb) for hydrogen storage. *Int. J. Hydrogen Energy* **2004**, *29*, 73–80. [CrossRef]
4. Ul Haq, B.; Kanoun, M.B.; Ahmed, R.; Bououdina, M.; Goumri-Said, S. Hybrid functional calculations of potential hydrogen storage material: Complex dimagnesium iron hydride. *Int. J. Hydrogen Energy* **2014**, *39*, 9709–9717. [CrossRef]
5. Kumar, D.; Singh, A.; Prasad Tiwari, G.; Kojima, Y.; Kain, V. Thermodynamics and kinetics of nano-engineered Mg-MgH_2 system for reversible hydrogen storage application. *Thermochim. Acta* **2017**, *652*, 103–108. [CrossRef]
6. Trivedi, D.R.; Bandyopadhyay, D. Study of adsorption and dissociation process of H_2 molecule on MgnRh clusters: A density functional investigation. *Int. J. Hydrogen Energy* **2016**, *41*, 20113–20121. [CrossRef]
7. Song, Y.; Guo, Z.X.; Yang, R. Influence of selected alloying elements on the stability of magnesium dihydride for hydrogen storage applications: A first-principles investigation. *Phys. Rev. B* **2004**, *69*, 094205. [CrossRef]
8. Vajeeston, P.; Ravindran, P.; Kjekshus, A.; Fjellvåg, H. Pressure-Induced Structural Transitions in MgH_2. *Phys. Rev. Lett.* **2002**, *89*, 175506. [CrossRef] [PubMed]
9. Kurko, S.; Matović, L.; Novaković, N.; Matović, B.; Jovanović, Z.; Mamula, B.P.; Grbović Novaković, J. Changes of hydrogen storage properties of MgH_2 induced by boron ion irradiation. *Int. J. Hydrogen Energy* **2011**, *36*, 1184–1189. [CrossRef]
10. Song, M.Y.; Kwon, S.N.; Park, H.R.; Hong, S.-H. Improvement in the hydrogen storage properties of Mg by mechanical grinding with Ni, Fe and V under H_2 atmosphere. *Int. J. Hydrogen Energy* **2011**, *36*, 13587–13594. [CrossRef]
11. Noritake, T.; Aoki, M.; Towata, S.; Seno, Y.; Hirose, Y.; Nishibori, E.; Takata, M.; Sakata, M. Chemical bonding of hydrogen in MgH_2. *Appl. Phys. Lett.* **2002**, *81*, 2008–2010. [CrossRef]

12. Liang, G.; Huot, J.; Boily, S.; Van Neste, A.; Schulz, R. Catalytic effect of transition metals on hydrogen sorption in nanocrystalline ball milled MgH$_2$ – Tm (Tm = Ti, V, Mn, Fe and Ni) systems. *J. Alloys Compd.* **1999**, *292*, 247–252. [CrossRef]

13. Shang, C.X.; Bououdina, M.; Guo, Z.X. Structural stability of mechanically alloyed (Mg + 10Nb) and (MgH$_2$ + 10Nb) powder mixtures. *J. Alloys Compd.* **2003**, *349*, 217–223. [CrossRef]

14. Rivoirard, S.; de Rango, P.; Fruchart, D.; Charbonnier, J.; Vempaire, D. Catalytic effect of additives on the hydrogen absorption properties of nano-crystalline MgH$_2$(X) composites. *J. Alloys Compd.* **2003**, *356*, 622–625. [CrossRef]

15. Oelerich, W.; Klassen, T.; Bormann, R. Metal oxides as catalysts for improved hydrogen sorption in nanocrystalline Mg-based materials. *J. Alloys Compd.* **2001**, *315*, 237–242. [CrossRef]

16. Aguey-Zinsou, K.F.; Ares Fernandez, J.R.; Klassen, T.; Bormann, R. Effect of Nb$_2$O$_5$ on MgH$_2$ properties during mechanical milling. *Int. J. Hydrogen Energy* **2007**, *32*, 2400–2407. [CrossRef]

17. Song, M.; Bobet, J.-L.; Darriet, B. Improvement in hydrogen sorption properties of Mg by reactive mechanical grinding with Cr$_2$O$_3$, Al$_2$O$_3$ and CeO$_2$. *J. Alloys Compd.* **2002**, *340*, 256–262. [CrossRef]

18. Kresse, G.; Furthmüller, J. Efficient Iterative Schemes for Ab Initio Total-Energy Calculations Using a Plane-Wave Basis Set. *Phys. Rev. B Condens. Matter* **1996**, *54*, 11169–11186. [CrossRef] [PubMed]

19. Kresse, G.; Joubert, D. From ultrasoft pseudopotentials to the projector augmented-wave method. *Phys. Rev. B* **1999**, *59*, 1758–1775. [CrossRef]

20. Perdew, J.P.; Burke, K.; Ernzerhof, M. Generalized Gradient Approximation Made Simple. *Phys. Rev. Lett.* **1996**, *77*, 3865–3868. [CrossRef] [PubMed]

21. Hammer, B.; Hansen, L.B.; Nørskov, J.K. Improved adsorption energetics within density-functional theory using revised Perdew-Burke-Ernzerhof functionals. *Phys. Rev. B* **1999**, *59*, 7413–7421. [CrossRef]

22. Monkhorst, H.J.; Pack, J.D. Special points for Brillouin-zone integrations. *Phys. Rev. B* **1976**, *13*, 5188–5192. [CrossRef]

23. Perdew, J.P.; Zunger, A. Self-interaction correction to density-functional approximations for many-electron systems. *Phys. Rev. B* **1981**, *23*, 5048–5079. [CrossRef]

24. Heyd, J.; Scuseria, G.E.; Ernzerhof, M. Erratum: Hybrid functionals based on a screened Coulomb potential. *J. Chem. Phys.* **2006**, *118*, 8207. [CrossRef]

25. Togo, A.; Oba, F.; Tanaka, I. First-Principles Calculations of the Ferroelastic Transition Between Rutile-Type and CaCl$_2$-Type SiO$_2$ at High Pressures. *Phys. Rev. B Condens. Matter* **2008**, *78*. [CrossRef]

26. Gonze, X.; Lee, C. Dynamical matrices, Born effective charges, dielectric permittivity tensors, and interatomic force constants from density-functional perturbation theory. *Phys. Rev. B* **1997**, *55*, 10355–10368. [CrossRef]

27. Baroni, S.; de Gironcoli, S.; Dal Corso, A.; Giannozzi, P. Phonons and related crystal properties from density-functional perturbation theory. *Rev. Mod. Phys.* **2001**, *73*, 515–562. [CrossRef]

28. Gu, T.; Wang, Z.; Tada, T.; Watanabe, S. First-principles simulations on bulk Ta$_2$O$_5$ and Cu/Ta$_2$O$_5$/Pt heterojunction: Electronic structures and transport properties. *J. Appl. Phys.* **2009**, *106*, 262907. [CrossRef]

29. Sun, R.; Wang, Z.; Saito, M.; Shibata, N.; Ikuhara, Y. Atomistic mechanisms of nonstoichiometry-induced twin boundary structural transformation in titanium dioxide. *Nat. Commun.* **2011**, *6*, 7120. [CrossRef] [PubMed]

30. Wang, Z.; Saito, M.; Mckenna, K.P.; Gu, L.; Tsukimoto, S.; Shluger, A.L.; Ikuhara, Y. Atom-resolved imaging of ordered defect superstructures at individual grain boundaries. *Nature* **2011**, *479*, 380–383. [CrossRef] [PubMed]

31. McKenna, K.P.; Hofer, F.; Gilks, D.; Lazarov, V.K.; Chen, C.; Wang, Z.; Ikuhara, Y. Atomic-scale structure and properties of highly stable antiphase boundary defects in Fe$_3$O$_4$. *Nat. Commun.* **2014**, *5*, 5740. [CrossRef] [PubMed]

32. Wang, Z.; Saito, M.; Mckenna, K.P.; Fukami, S.; Sato, H.; Ikeda, S.; Ohno, H.; Ikuhara, Y. Atomic-Scale Structure and Local Chemistry of CoFeB-MgO Magnetic Tunnel Junctions. *Nano Lett.* **2016**, *16*, 1530–1536. [CrossRef] [PubMed]

33. Yan, H.; Ziyu, H.; Xu, G.; Xiaohong, S. Structural, electronic and photocatalytic properties of atomic defective BiI3 monolayers. *Chem. Phys. Lett.* **2018**, *691*, 341–346. [CrossRef]

34. García, G.N.; Abriata, J.P.; Sofo, J.O. Calculation of the electronic and structural properties of cubic Mg$_2$NiH$_4$. *Phys. Rev. B* **1999**, *59*, 11746–11754. [CrossRef]

35. Chen, Y.; Dai, J.; Xie, R.; Song, Y.; Bououdina, M. First principles study of dehydrogenation properties of alkali/alkali-earth metal doped Mg_7TiH_{16}. *J. Alloys Compd.* **2017**, *728*, 1016–1022. [CrossRef]
36. Shelyapina, M.G.; Fruchart, D.; Wolfers, P. Electronic structure and stability of new FCC magnesium hydrides Mg_7MH_{16} and Mg_6MH_{16} (M = Ti, V, Nb): An ab initio study. *Int. J. Hydrogen Energy* **2010**, *35*, 2025–2032. [CrossRef]
37. Dai, J.H.; Song, Y.; Yang, R. First Principles Study on Hydrogen Desorption from a Metal (=Al, Ti, Mn, Ni) Doped MgH_2 (110) Surface. *J. Phys. Chem. C* **2010**, *114*, 11328–11334. [CrossRef]
38. Kumar, M.; Kamal, R.; Thapa, R. Screening based approach and dehydrogenation kinetics for MgH_2: Guide to find suitable dopant using first-principles approach OPEN. *Sci. Rep.* **2017**, *7*, 15550. [CrossRef] [PubMed]
39. Van Mal, H.H.; Buschow, K.H.J.; Miedema, A.R. Hydrogen absorption in $LaNi_5$ and related compounds: Experimental observations and their explanation. *J. Less Common Met.* **1974**, *35*, 65–76. [CrossRef]
40. Lakhal, M.; Bhihi, M.; Benyoussef, A., El Kenz, A.; Loulidi, M ; Naji, S. The hydrogen ab/desorption kinetic properties of doped magnesium hydride MgH_2 systems by first principles calculations and kinetic Monte Carlo simulations. *Int. J. Hydrogen Energy* **2015**, *40*, 6137–6144. [CrossRef]
41. Alapati, S.V.; Johnson, J.K.; Sholl, D.S. Identification of Destabilized Metal Hydrides for Hydrogen Storage Using First Principles Calculations. *J. Phys. Chem. B* **2006**, *110*, 8769–8776. [CrossRef] [PubMed]
42. Yu, R.; Lam, P.K. Electronic and structural properties of MgH2. *Phys. Rev. B* **1988**, *37*, 8730–8737. [CrossRef]
43. Westerwaal, R.J.; Broedersz, C.P.; Gremaud, R.; Slaman, M.; Borgschulte, A.; Lohstroh, W.; Tschersich, K.G.; Fleischhauer, H.P.; Dam, B.; Griessen, R. Study of the hydride forming process of in-situ grown MgH_2 thin films by activated reactive evaporation. *Thin Solid Films* **2008**, *516*, 4351–4359. [CrossRef]

MDPI

St. Alban-Anlage 66

4052 Basel

Switzerland

Tel. +41 61 683 77 34

Fax +41 61 302 89 18

www.mdpi.com

Metals Editorial Office

E-mail: metals@mdpi.com

www.mdpi.com/journal/metals